Knight Ora Willis,   United Onthologists of Maine

A List of the Birds of Maine

Showing their Distribution by Counties and their Status in Each County

Knight Ora Willis, United Onthologists of Maine

A List of the Birds of Maine
Showing their Distribution by Counties and their Status in Each County

ISBN/EAN: 9783744648363

Printed in Europe, USA, Canada, Australia, Japan

Cover: Foto ©berggeist007 / pixelio.de

More available books at **www.hansebooks.com**

BULLETIN   No. 3.

THE UNIVERSITY OF MAINE

DEPARTMENT OF NATURAL HISTORY.

A LIST OF THE

BIRDS OF MAINE

Showing their Distribution by Counties

And their Status in Each County.

Prepared under the auspices of the United Ornithologists of Maine

BY

ORA W. KNIGHT, B. S.,

Assistant in Natural History.

AUGUSTA
KENNEBEC JOURNAL PRINT
1897

STATE OF MAINE.

ORONO, ME., March 28, 1897.

President A. W. Harris:

I take pleasure in submitting herewith, Bulletin No. 3, from the Laboratory of Natural History. This Bulletin, prepared by Mr. O. W. Knight of the Class of 1895, and until recently assistant in this department, is a valuable contribution to the Ornithology of Maine, and will be an honor both to the editor and to the institution.

Respectfully,

FRANCIS L. HARVEY,

Professor of Natural History.

INTRODUCTION.

As all the previous lists of Maine birds are inaccessible to the majority of Ornithologists, it has seemed desirable to publish one in which our present knowledge of the distribution and status of the birds of the state may be given. Many records taken from the various Ornithological publications have been incorporated in the list.

While it has seemed desirable to make this an annotated list, the chief aim of the notes is to point out the distribution and status of each species in the various counties of the state. Remarks regarding the habits, songs and migration dates of many birds are also included.

No record has been admitted except upon the best of evidence as to its authenticity. The distribution and relative abundance of each species have been reported on by one or more observers from each county, except Lincoln which is represented only by a few records taken from material which has already been published. The members of the United Ornithologists of Maine have rendered aid by furnishing many notes of interest, and their efforts have been heartily seconded by many other prominent Ornithologists of the state.

The classification and nomenclature of The American Ornithologists' Union have been followed. A summary of the status and abundance of each species is followed by the county records. The counties are arranged in alphabetical order, and following each is a brief statement of the bird's status in that county, and the name of the observer who is responsible for the statement.

All remarks regarding the habits, abundance, migration dates, etc., of individual species apply only to this state, unless expressly stated otherwise in the text.

Maine is of great Ornithological interest owing to its being the most eastern meeting point of the Canadian and Alleghanian faunæ.

Many birds belonging to warmer regions straggle across our southern and western boundaries, while many northern birds likewise occur in the northern and eastern counties.

A partial bibliography has been appended, but notices regarding many of the publications cited have been taken from other publications, and consequently the author is not responsible for any errors in titles which may occur.

Species which have been reported in previous lists upon insufficient evidence, those which have been taken near our boundaries, and those which may reasonably be expected to be detected in the state at some future date are given in a hypothetical list.

A supplement to the list will be issued as soon as enough additional information to warrant it has been secured, and all persons having knowledge of the occurrence in the state of any species not cited are requested to send notice of the same to the author. It is also hoped that any mistakes which may occur will be pointed out for correction.

Below is given a list of the counties of the state with the names of the observers from each county.

Androscoggin : E. E. Johnson lists 167 species taken or seen near Lewiston ; A. R. Pike reports on 66 of the rarer birds from the same locality ; Prof. A. B. Call has furnished notes on 105 species likewise observed near Lewiston ; these notes are supplemented by records from "The Birds of Androscoggin County," and information received from Prof. J. Y. Stanton of Bates College.

Aroostook : The majority of the records from this county are credited to Mr. Chas. F. Batchelder, and were taken from a series of articles entitled "Notes on the Summer Birds of the Upper St. John" which were published in early numbers of the Bulletin of the Nuttall Ornithological Club ; notes on certain species observed by O. W. Knight at Sherman, Fort Fairfield, Houlton, Caribou, Limestone and Presque Isle, have also been used.

Cumberland : J. C. Mead lists 147 species from the northern part of the county ; these are supplemented by notes from A. H. Norton and Dr. H. H. Brock, and extracts from Brown's "Catalogue of the Birds of Portland."

Franklin : J. Merton Swain has given information regarding 130 species from the southern part of the county, and F. M. Richards reports on 147 from the same section ; these are supplemented by a few extracts from notes furnished by W. A. Lee & W. E. McLain, and information received from E. F. Cobb, regarding birds found at New Vineyard and Rangeley respectively.

Hancock: E. F. Murch lists 70 species found in the vicinity of Ellsworth, and A. G. Dorr of Bucksport reports 103 from his locality; O. W. Knight gives notes regarding many of the seabirds which are found breeding among the islands along the coast.

Kennebec: The Gardiner Branch of the United Ornithologists of Maine has reported on the birds of this county; the records of the rarer species are credited to the individuals responsible for them, while the others are credited to the Branch; the members of which are, William L. Powers, Clara M. Burleigh, Sadie M. Jewett, Mallian Reed, Lillian Holmes, Martha Webster, Lizzie Church, Austin P. Larrabee, Homer Dill, Maurice Royal, Fred Dill, Ralph Hunt, George Royal, Harold Peacock, Lincoln Harlow, Harold Giddings, Ralph Harden, Charles Austin, George Dow, Edmund Welch, Eben Haley, L. M. Sanborn and L. W. Robbins; additional records are taken from Prof. Hamlin's "List of the Birds of Waterville."

Knox: Fred Rackliff of Spruce Head lists 205 species from the county, and A. H. Norton adds 10 more; it is probable that some of the former's notes regarding the presence of certain sandpipers and allied birds in summer, refer to their return from the north in late July and early August, and do not imply that they are present throughout the entire summer.

Lincoln: This county is represented by a few records taken from various publications, no observer having been found who was willing to furnish a county list. We may safely say that the birds of this small county are identical with those found in Knox on the east and Sagadahoc on the west.

Oxford: J. Waldo Nash of Norway lists 164 species, and E. E. Johnson reports on 89 found near East Hebron; a few notes from Maynard's "List of the Birds of Coos County, New Hampshire, and Oxford County, Maine," are used as are some from Verrill's "List of the Birds of Norway," (Maine).

Penobscot: O. W. Knight gives an annotated list of 155 species found in the southern part of the county, and Manly Hardy adds notes on 24 of the rarer birds; Granville Gray of Oldtown adds 4 species to the list.

Piscataquis : Wallace Homer submits a fully annotated list of 116 species, and Charles Whitman reports on 56.

Sagadahoc : Herbert L. Spinney has found 180 species here and his list is copiously annotated; C. C. Spratt of North Bridgton, lists 79 species.

Somerset : C. H. Morrell presents a fully annotated list of 145 species from the southern part of the county, while Prof. F. L. Harvey and O. W. Knight add a few notes from Jackman in the in the northern part.

Waldo : C. C. Spratt gives a list of 80 species ; O. W. Knight supplements this by notes on many of the birds observed by him within the county.

Washington : George A. Boardman of Calais has observed and taken 257 species within Washington County. His list is copiously annotated and is the result of long years of careful observation.

York : Stephen J. Adams submits a list embracing 102 species which he has observed near the towns of Hiram, Oxford County, Cornish, York County, and Baldwin, Cumberland County; his notes are thus referable to three counties although in the list they have been credited as noted here : Charles S. Butters has given notes regarding a few seabirds taken at Biddeford Pool.

Other matters being equal, the first notes received have been used to set forth the county status of a given species, and subsequent notes on the same species from observers in the same county have not been used unless the conclusions set forth are somewhat different. In some cases the notes of different observers in a given county have been used when they indicate a difference in the status or abundance of a species in the different parts of the county reported on by the observers.

Thanks are due to the University of Maine and to Professor F. L. Harvey for allowing me free access to its collections and library ; to William Brewster of Cambridge, Mass., for identifying specimens submitted to him ; to Harry Merrill of Bangor for the loan of various publications germain to the list ; to J. C. Mead of North Bridgton, Arthur H. Norton of Westbrook, and Prof. William L. Powers of Gardiner for the aid they have rendered ; to these and many others who have helped on the work in hand cordial thanks are extended.

ORA W. KNIGHT.

BANGOR, March 28, 1897.

LAWS OF THE STATE OF MAINE WHICH RELATE ESPECIALLY TO ORNITHOLOGY OR OÖLOGY.

(Taken from the Fish and Game Laws.)

1895, c. 125, § 21. Whoever kills or has in his possession, except alive, or exposes for sale, any wood duck, dusky duck, commonly called black duck, teal or grey duck, between the first days of May and September, or kills, sells, or has in his possession, except alive, any ruffed grouse, commonly called partridge, between the first days of December and September 20th, or woodcock, between the first days of December and September following; or kills, sells or has in his possession, except alive, any quail between the first day of December and the first day of October following, or pinnated grouse, commonly called prairie chicken, between the first days of January and September, or plover between the first days of May and August, forfeits not less than $5 nor more than $10, for each bird so killed, had in possession or exposed for sale. And no person shall at any one time, kill, expose for sale, or have in possession, except alive, more than 30 of each variety of birds above named, during the respective open seasons, nor shall any person at any time kill, expose for sale, or have in possession, except alive, any of the above named varieties of birds except for consumption within this state, under a penalty of $5 for each bird so unlawfully killed, exposed for sale or in possession; nor shall any person or corporation carry or transport from place to place in open season any of the above mentioned birds unless open to view, tagged and plainly labeled with the owner's name, and accompanied by him, under the same penalty; any person, not the actual owner of such birds, who, to aid another in such transportation falsely represents himself to be the owner thereof, shall be liable to the same penalty; nor shall any person or corporation carry or transport at any one time more than 15 of any one variety of birds above named, as the property of one man under the same penalty; nothing in this section shall prevent any marketman or

provision dealer having an established place of business in this
state, from purchasing at his place of business, any bird lawfully
caught, killed or destroyed, or any part thereof, and selling the
same in open season at retail to his local customers.

1889, c. 248 § 22. Whoever, at any time or in any place, with
any trap, net, snare, device or contrivance, other than the usual
method of sporting with fire-arms, takes wild duck of any variety,
quail, grouse, partridge or woodcock, forfeits five dollars for each
bird so taken.

1889, c. 249, § 23. Whoever, kills or has in his possession,
except alive, any birds commonly known as larks, robins, swallows,
sparrows or orioles, or other insectivorous birds, crows, English
sparrows, and hawks excepted, forfeits not less than one dollar,
nor more than five dollars, for each such bird killed, and the possession by any person of such dead bird, is prima facie evidence that
he killed such bird.

R. S., c. 30, § 24. Whoever at any time wantonly takes or
destroys the nest, eggs, or unfledged young of any wild bird except
crows, hawks and owls, or takes any eggs or young from such
nests, except for the purpose of preserving the same as specimens,
or of rearing said young alive, forfeits not less than one dollar nor
more than ten dollars for each nest, egg, or young so taken or
destroyed.

TRANSPORTATION.

R. S., c. 30, § 25. Whoever carries or transports from place to
place, any of the birds named herein, during the period in which
the killing of such bird is prohibited, forfeits five dollars for each
bird so carried or transported.

PROTECTION OF CAPERCAILIZE AND OTHER BIRDS.

1895, c. 149, § 1. It shall be unlawful for a term of five years
to hunt for, take, catch, kill or destroy any of the following named
birds under a penalty of fifty dollars for the offense, and twenty-
five dollars for every bird so taken, caught, killed or destroyed.
The capercailzie, or cock of the woods, so called, black game, so
called, or any species of the pheasant, except the partridge, so
called.

SECT. 2. All fines and penalties under this act, shall be enforced
in the same manner, as for the violation of laws relating to the
illegal killing of game.

EVERY SUNDAY IS CLOSE TIME.

R. S., c. 30, § 27. Sunday is a close time, on which it is not lawful to hunt, kill or destroy game or birds of any kind, under the penalties imposed therefor during other close times; but the penalties already imposed for violation of the Sunday laws are not repealed or diminished.

TAXIDERMIST.

1895, c. 50, § 1. The commissioners of inland fisheries and game may, upon application, issue a license to such persons as taxidermists. who, in their judgment, are skilled in that art, of good reputation and friendly to the fish and game laws of this state. For such license the applicant shall pay into the state treasury the sum of $5, to be credited as additional to the funds appropriated by the state to inland fisheries and game, and be in force for three years from the date of its issue, unless sooner revoked. Such licensee may at all times have in his possession at his place of business, fish and game, or parts thereof, lawfully caught or killed in open time for the sole purpose of preparing for, and mounting the same; and such fish and game or parts thereof may be transported to such licensee and retained by him for the purposes aforesaid, under such rules, restrictions and limitations as shall, from time to time be made by said commissioners and stated in such original license and additions made thereto from time to time by said commissioners.

SECT. 2. Such licenses may be revoked by said commissioners at any time after notice and an opportunity for a hearing; and every licensee and carrier violating any of the provisions of this act, or of the rules, restrictions or limitations set out in said license and additions thereto, shall, on complaint before any trial justice or municipal or police court, be fined not less than $20 nor more than $50.

CERTAIN PERSONS AUTHORIZED TO TAKE BIRDS AND THEIR NESTS AND EGGS FOR SCIENTIFIC PURPOSES.

1885, c. 333, § 1. Upon the request and recommendation of the fish and game commissioners, the governor, with the advice and consent of the council may commission persons to take, kill, capture and have in possession any species of bird other than domestic, and the nests and eggs thereof for scientific purposes ; but the number of commissions in force shall not exceed ten at any time.

SECT. 2. No person to whom such commission may be granted, shall sell, offer for sale, or take any compensation for specimens of birds, nests or eggs, nor dispose of the same by gift or otherwise, to be taken from the State, except by exchange of specimens for scientific purposes ; and for any violation of any of the provisions of this section, such person shall be subject to a fine of not less than ten nor more than fifty dollars, to be recovered by complaint before any trial justice or municipal judge.

THE BIRDS OF MAINE.

Order PYGOPODES. Diving Birds.

Suborder PODICIPEDES. Grebes.

Family PODICIPIDÆ. Grebes.

Genus COLYMBUS. Linnæus.

Subgenus COLYMBUS.

1. (2).* Colymbus holbœllii (*Reinh.*). Holbœll's Grebe.

Occurs along the coast as a winter resident in limited numbers, September to late April. There are no records from the interior counties of the state. It breeds north of our limits.

County Records.—Cumberland, "rather uncommon winter resident" (Brown's Cat. Birds of Portland, p. 36); Knox, "migrant" (Norton); Sagadahoc, "not common, few in winter" (Spinney); Washington, "common" (Boardman); York, (Butters).

Subgenus DYTES Kaup.

2. (3). Colymbus auritus *Linn.* Horned Grebe.

Common as a migrant along the coast and to a less extent in the interior; a few are also found in winter along the coast. It is a rare summer resident in our northeastern counties.

County Records.—Cumberland, "rare" (Mead); "rather common in migrations, a few probably winter" (Brown's Cat. Birds of Portland, p. 36); Hancock, "common migrant" (Dorr); Knox, "migrant" (Norton); Oxford, "very rare" (Nash); Penobscot, "one shot at East Orrington" (Hardy); Piscataquis, "rare" (Homer); Sagadahoc, "common December to April" (Spinney); Washington, "common, a few breed" (Boardman); York, (Butters).

---

* The numbers at the left are the Maine numbers and those in parenthesis are the A. O. U. numbers of the species.

Genus PODILYMBUS Lesson.

3.　(6).　Podilymbus podiceps (*Linn.*).　Pied-billed Grebe.

Commonest in migrations, but also quite a common summer resident on various bodies of fresh water throughout the state. In migrations it is commoner along the coast than in the interior.

County Records.—Audroscoggin, "common summer resident" (Johnson); Aroostook, "rare, breeds" (Batchelder, Bull. Nutt. Orn. Club, Vol. 7, p. 152); Cumberland, "common" (Mead); Franklin, "I am informed by Mr. Elmer Cobb that he has taken eggs at Rangeley Lake" (Knight); Hancock, "common" (Dorr); Kennebec, "rare summer resident" (Robbins); Knox, (Rackliff); Penobscot, "summer resident" (Knight); Sagadahoc, "not common, seen only in fall" (Spinney); Somerset, "common summer resident" (Morrell); Washington, "common, few breed" (Boardman).

Suborder CEPPHI.　Loons and Auks.

Family URINATORIDÆ.　Loons.

Genus URINATOR Cuvier.

4.　(7).　Urinator imber (*Gunn.*).　Loon.

A resident along the coast, a common summer resident and breeder on the ponds and lakes of the interior. It does not breed on the seacoast, although birds may be seen there all summer. These are probably immature or sterile individuals.

County Records.—Androscoggin, "common summer resident" (Johnson); Cumberland, "summer resident" (Mead); Franklin, "common summer resident" (Swain); Hancock, "common resident" (Dorr); Kennebec, "rare summer resident" (Gardiner Branch); Knox "resident" (Rackliff); Oxford, "breeds commonly" (Nash); Penobscot, "summer resident, not so common as formerly" (Knight); Piscataquis, "common, breeds" (Homer); Sagadahoc, "common resident" (Spinney); Somerset, "not very common summer resident" (Morrell); Waldo, (Spratt); Washington, "common, breeds" (Boardman); York, "common summer resident" (Adams).

5.　(11).　Urinator lumme (*Gunn.*).　Red-throated Loon.

Quite a common fall and spring migrant along the coast and on the ponds and lakes. It is also a winter resident coastwise. It breeds in high latitudes.

County Records.—Cumberland, "common in migration" (Brown's Cat. Birds of Portland, p. 36); Kennebec, "accidental" (Dill); Knox, "migrant" (Rackliff); Penobscot, "immature birds often" (Hardy); Sagadahoc, "from late fall to spring" (Spinney); Somerset, "Mr. C. W. Savage, the postmaster at Flagstaff, has a bird which from the description and measurements must be this species, and which was shot at Flagstaff pond in the fall of 1896" (Knight); Washington, "common" (Boardman); York, (Butters).

Family ALCIDÆ. Auks, Murres, and Puffins.
Subfamily FRATERCULINÆ. Puffins.
Genus LUNDA Pallas.

6. (12). Lunda cirrhata *Pall.* Tufted Puffin.

This is a Pacific coast bird which is of accidental occurrence on our coast, and of which there is only one specimen recorded. (Cf Allen, The Auk, Vol. 2, p. 388). This record is based upon the authority of Audubon, who has stated that one of this species was taken at the mouth of the Kennebec River in the winter of 1831-'32.

Genus FRATERCULA Brisson.

7. (13). Fratercula arctica (*Linn.*). Puffin.

Occurs commonly as a winter visitor to our coast, but is said to have nested in limited numbers on Seal Island as late as 1888, and it is reported that some six pairs of these birds nested on Matinicus Rock as late as the summer of 1896. It is only a question of a year or so when this species will cease to nest along our coast.

County Records.—Cumberland, "not common winter visitant"(Brown's Cat. Birds of Portland, p. 36); Hancock, "winter visitor" (Dorr); Knox, "resident"* (Rackliff); Sagadahoc, "only in winter, quite scarce" (Spinney); Washington, "winter visitant; a few breed at Grand Menan, N. B." (Boardman).

Subfamily PHALERINÆ. Auklets, Murrelets, Guillemots.
Genus CEPPHUS Pallas.

8. (27). Cepphus grylle (*Linn.*). Black Guillemot.

A common resident and breeder along the coast from Knox County eastward, while elsewhere it occurs only as a winter visitant.

---

*Only a very few of these birds remain through the summer although the fact that some do remain would justify its being cited as a rare resident (Editor).

County Records. — Cumberland, "not uncommon winter visitant" (Brown's Cat. Birds of Poitland, p. 36); Franklin, "accidental" (Richards); Hancock, "very common resident about the outer islands" (Knight); Knox, "resident" (Racklifl); Penobscot, "taken in Brewer in winter" (Hardy); Sagadahoc, "plenty in winter" (Spinney); Washington, "resident" (Boardman).

Subfamily ALCINÆ.   Auks and Murres.

Genus URIA Brisson.

9. (30).   Uria troile (*Linn.*).   Murre.

Winter visitor in limited numbers along the coast.   In most cases where this species has been reported a careful investigation has brought out the fact that the reports should have referred to the succeeding species.

County Records.—Cumberland, "rare winter visitor" (Brown's Cat. Birds of Portland, p. 36); Knox, "rare winter visitant" (Rackliff); Washington, (Boardman).

10. (31).   Uria lomvia (*Linn.*).   Brünnich's Murre.

A quite common winter visitor along the coast, especially on the outer islands.   It is reported by Mr. Boardman as breeding on the islands near Grand Menan, New Brunswick, in limited numbers and in company with the preceding species.

County Records.—Cumberland, "not uncommon winter visitant" (Brown's Cat. Birds of Portland, p. 36); Hancock, "winter visitor" (Knight); Knox, "winter visitor" (Rackliff); Sagadahoc, "common in winter" (Spinney); Somerset, "found one dead on the ice, near Pittsfield, December 31, 1896" (Morrell); Washington, (Boardman).

Genus ALCA Linnæus.

11. (32).   Alca torda (*Linn.*).   Razor-billed Auk.

Quite a common winter visitor to the outer islands of the coast. Mr. Boardman reports that a few still breed at Grand Menan, New Brunswick.

County Records.—Cumberland, "a winter visitor of quite frequent occurrence" (Brown's Cat. Birds of Portland, p. 36); Knox, "winter visitor" (Rackliff); Sagadahoc, "common in winter" (Spinney); Washington, "winter" (Boardman).

Genus Plautus Brünnich.

12. (33). Plautus impennis (*Linn.*). Great Auk.

This species has become extinct through the agency of man, during the present century. It was probably found along the entire coast of Maine, in winter at least. Evidences of this birds having once occurred at Gouldsborough, Hancock County, are recorded in the Report of the Maine Board of Agriculture for 1877, p. 261.

Subfamily ALLINÆ. Dovekies.

Genus ALLE Link.

13. (34). Alle alle (*Linn.*). Dovekie.

A not uncommon visitor some winters along the coast; sometimes driven inland by severe storms. It breeds in high latitudes.

County Records. — Cumberland, "rather irregular winter visitor" (Brown's Cat. Birds of Portland, p. 36); Knox, "winter visitor" (Racklift); Penobscot, "have seen several taken near Brewer" (Hardy); Sagadahoc, "in winter, not plenty" (Spinney); Washington, "in winter only" (Boardman).

Order LONGIPENNES. Long-winged Swimmers.

Family STERCORARIIDÆ. Skuas and Jaegers.

Genus STERCORARIUS Brisson.

14. (36). Stercorarius pomarinus (*Temm.*). Pomarine Jaeger.

Of quite regular occurrence in spring and fall; casual in summer. Nests in the far north.

County Records.—Cumberland, (Brown's Cat. Birds of Portland, p. 35); Knox, "have seen it in summer" (Norton); Sagadahoc, (Spinney); Washington, "not common" (Boardman); York, "I have a specimen from the town of York" (Norton).

15. (37). Stercorarius parasiticus (*Linn.*). Parasitic Jaeger.

Quite common off shore in spring and fall; nests in the far north.

County Records.—Cumberland, "said to be not uncommon off shore" (Brown's Cat. Birds of Portland, p. 35); Knox, "migrant" (Racklift); Sagadahoc, "not common near shore" (Spinney); Washington, "rare" (Boardman).

2

16. (38). Stercorarius longicaudus *Vieill.* Long-tailed Jaeger.

This bird occurs in a manner precisely the same as the two preceding species. It is, however, seemingly not so abundant as it is not reported from as many counties as the other two are.

County Records.—Sagadahoc, "not common" (Spinney); Washington, "common in fall" (Boardman).

Family LARIDÆ. Gulls and Terns.

Subfamily LARINÆ. Gulls.

Genus RISSA Stephens.

17. (40). Rissa tridactyla (*Linn.*). Kittiwake.

Common winter visitor along the coast; breeds in Arctic regions.

County Records.—Cumberland, "common winter resident" (Brown's Cat. Birds of Portland, p. 34); Hancock, "rare in winter" (Dorr); Knox, "winter visitor" (Racklift); Oxford, "one in October, 1890, on Lovewell's Pond" (Nash); Sagadahoc, "plenty from November to April" (Spinney); Washington, "abundant fall migrant" (Boardman); York, (Butters).

Genus LARUS Linnæus.

18. (42). Larus glaucus *Brünn.* Glaucous Gull.

A rare winter visitor to our coast; breeds in Arctic regions.

County Records.—Cumberland, "a specimen was shot at Peake's Island, April 27, 1883" (Brown's Cat. Birds of Portland, p. 37); Hancock, "rare in winter" (Dorr); Knox, "winter" (Racklift); Washington, "rare, winter only" (Boardman).

19. (43). Larus leucopterus *Faber.* Iceland Gull.

This species is probably of much commoner occurrence along the coast in winter than one would imagine from the paucity of reports concerning it.

County Records.—Cumberland, "Audubon found it in Portland Harbor" (Orn. Biog. III, p. 553); Hancock, "West Sullivan" (Brewster, Bull. Nutt. Orn. Club, Vol. 8, p. 251); Washington, "winter only"(Boardman).

20. (47). Larus marinus *Linn.* Great Black-backed Gull.

Of common occurrence along the coast in spring, autumn and winter, September to March. A few stragglers are occasionally seen in summer, but it is not known to breed in the state. Mr. Boardman states that a few breed about Grand Menan, New Brunswick.

County Records.—Cumberland, "common winter resident" (Brown's Cat. Birds of Portland, p. 34); Hancock, "rare" (Dorr); Knox, "winter visitor" (Rackliff); Sagadahoc, "few in summer, plenty in winter" (Spinney); Washington, "winter" (Boardman).

21. (51). **Larus argentatus** *Brünn.* Herring Gull.

An accidental visitor to our coast, but one specimen known. Mr. Fred Rackliff of Spruce Head, Knox County, has a Gull taken in that county which is unquestionably referable to this species. This record is verified by Mr. A. H. Norton.

22. (51a). **Larus argentatus smithsonianus** *Coues.* American Herring Gull.

A very common resident along the seacoast, and quite a common summer resident on the larger lakes of the interior. A large colony of these birds nest on an island of Moosehead Lake. In summer the eggs of the species form a staple article of food among the fisher. men along the coast, and, in consequence of their being constantly robbed, fresh eggs are often found late in August.

County Records.—Androscoggin, "fairly common migrant" (Johnson); Aroostook, "common on lakes and breeds" (Batchelder in Bull. Nutt. Orn. Club, Vol. 7, p. 152); Cumberland; "common spring visitor near Bridgton" (Mead), "resident throughout the year" (Brown's Cat. Birds of Portland, p. 34); Franklin, "rare, accidental" (Swain); Hancock, "common resident" (Knight); Kennebec, "rare" (Gardiner Branch); Knox, "resident" (Rackliff); Oxford, (given in Maynard's List, p. 30); Penobscot, "seen near Bangor in fall and early spring" (Knight); Piscataquis, "breeds on the lakes" (Homer); Sagadahoc, "few in summer, common in winter" (Spinney;) Somerset, "frequent visitor spring to fall" (Morrell); Waldo, "regular visitor throughout the year, does not breed within the county to my knowledge" (Knight); Washington, "common resident" (Boardman); York, (Butters).

23. (54). **Larus delawarensis** *Ord.* Ring-billed Gull.

Not very common as a spring and autumn migrant, but it is probably of more general occurrence than the meager county records would indicate.

County Records.—Cumberland, "quite common transient" (Brown's Cat. of Birds of Portland, p. 34); Hancock, "have one from near Bucksport" (Knight); Knox, "migrant" (Norton); Washington, "common in migration" (Boardman.)

24. (58). **Larus atricilla** *Linn.* Laughing Gull.

This species is of rare or irregular occurrence along the coast. It has been known to breed on the islands of Casco Bay.

County Records.—Cumberland, "quite common summer resident, breeding on the outer islands of Casco Bay" (Brown's Cat. Birds of Portland, p. 34); Knox, "summer" (Rackliff); Washington, "few about the islands in summer" (Boardman).

25. (60). Larus philadelphia (*Ord*). Bonaparte's Gull.

This species is very common along the coast in the migrations, and a few remain through the winter. It also occurs, though less commonly, on the larger lakes of the interior, but does not breed within the state. It is commonest from early August to November and from April to June.

County Records.—Androscoggin, (Pike); Cumberland, "not rare straggler" (Mead), "abundant during migrations" (Brown's Cat. Birds of Portland. p. 34); Hancock, "common in fall" (Dorr); Knox, "migrant" (Rackliff); Oxford, "very rare" (Nash); Piscataquis, "not uncommon migrant on the larger lakes" (Homer); Sagadahoc, "plenty from Nov. until April" (Spinney); Washington, "very abundant, none breed" (Boardman).

Genus XEMA Leach.

26. (62). Xema sabinii (*Sab.*). Sabine's Gull.

Of accidental occurrence within the state. One specimen is recorded in Smith's List of the Birds of Maine, and this same specimen is again recorded in the Bulletin of the Nuttall Ornithological Club, Vol. 3, page 195. Mr. Boardman took a specimen near Eastport, on Indian Island, New Brunswick, in May, 1878.

County Records.—Cumberland, "one taken in Scarborough on May 31st, 1877" (for this record see Forest and Stream, Vol. 20, page 205, under Smith's List of the Birds of Maine).

Subfamily STERNINÆ. Terns.

Genus GELOCHELIDON Brehm.

27. (63). Gelochelidon nilotica (*Hasselq.*). Gull-billed Tern.

This species has occurred as an accidental visitor from the south, several specimens having been taken at different times.

County Records.—Cumberland, "three taken in Sept. 1868, and birds also seen on May 21st, 1881" (Smith's List of the Birds of Maine, Forest and Stream, Vol. 20, p. 205); Washington, "accidental" (Boardman.)

Genus STERNA Linnæus.

Subgenus THALASSEUS Boie.

28. (64). Sterna tschegrava *Lepech.* Caspian Tern.

A rare but quite regular migrant along the coast. It is not known to breed within our limits.

County Records.—Cumberland, "a bird of the year taken at Richmond's Island in 1895" (Cf. Norton, Proc. Port. Soc. Nat. Histr., Apr. 1, 1897, p. 104); Knox, "very rare" (Racklift); Sagadahoc, "Aug. and Sept., not plenty" (Spinney); Washington, "seen in migrations, rare" (Boardman).

Subgenus STERNA Linnæus.

29. (70). Sterna hirundo *Linn.* Common Tern.

A very common summer resident along our coast, breeding in colonies on the various grassy islands, often in company with *Sterna paradisæa.* The eggs of both species are collected and eaten by the fishermen.

County Records.—Androscoggin, "rare migrant" (Johnson); Cumberland, "abundant summer resident" (Brown's Cat. Birds of Portland, p. 34); Franklin, "migrant" (Richards); Hancock, "breeds commonly on many of the islands" (Knight); Knox, "summer" (Racklift); Oxford, "very rare" (Nash); Penobscot, (Hardy); Sagadahoc, "breeds" (Spinney); Washington, "abundant" (Boardman); York, (Butters).

30. (71). Sterna paradisæa *Brünn.* Arctic Tern.

Summer resident along the coast. It is associated in the breeding season with the preceding species by which it is exceeded in point of numbers.

County Records.—Cumberland, "summer resident" (Brown's Cat. Birds of Portland, p. 34); Hancock, "summer resident, not common" (Knight); Knox, "summer" (Racklift); Piscataquis, "migrant about the lakes" (Homer); Sagadahoc, "breeds" (Spinney); Washington, "abundant" (Boardman).

31. (72). Sterna dougalli *Montag.* Roseate Tern.

Formerly a rare summer visitor to our southern coast, but has not been recorded of late years. Breeds sparingly on Sable Island, Nova Scotia, which is its northern limit. (Cf. Dwight, Memoirs of the Nutt. Orn. Club, No. II, "The Ipswich Sparrow," p. 16).

County Records.—Cumberland, "seen at Green Islands, Casco Bay" (Brewster, Bull. Nutt. Orn. Club, Vol. 4, p. 15); Knox, "formerly in summer, now exterminated" (Norton).

Subgenus STERNULA Boie.

32.   (74).   Sterna antillarum (*Less.*).   Least Tern.

Formerly of quite regular occurrence on the Green Islands in
Casco Bay.   It is reported by Mr. Boardman as being accidental
at Grand Menan, New Brunswick.   There are no recent records.

County Records.—Cumberland, "formerly occurred every year at Green
Islands, but none seen for a number of years" (Brown's Cat. Birds of
Portland, p. 35).

Subgenus HALIPLANA Wagler.

33.   (75).   Sterna fuliginosa *Gmel.*   Sooty Tern.

There is only one specimen recorded from the state, and this
was taken at Parkman in Piscataquis County, October 5, 1878.   It
is recorded by Ruthven Deane in the Bulletin of the Nuttall Ornithological Club, Vol. 5, page 64.

Genus HYDROCHELIDON Boie.

34.   (77).   Hydrochelidon nigra surinamensis (*Gmel.*).
Black Tern.

This species is of rare or casual occurrence in the migrations.
It is reported by Mr. Boardman as accidental at Grand Menan,
New Brunswick.

County Records.—Cumberland, "transient in autumn, uncommon"
(Brown's Cat. Birds of Portland, p. 35) ; York, "occurs at Wells Beach"
(Brown in Bull. Nutt. Orn. Club, Vol. 7, p. 190).

Family RYNCHOPIDÆ.   Skimmers.

Genus RYNCHOPS Linnæus.

35.   (80).   Rynchops nigra *Linn.*   Black Skimmer.

An accidental visitor from the south.   Several specimens have
been taken at various times.

County Records.—Knox, "Matinicus Island" (recorded in Smith's List
of Birds of Maine, Forest and Stream, Vol. 20, p. 205) ; Washington,
"accidental" (Boardman) ; York, "taken at Wells Bay" (recorded in
Smith's List as above).

Order TUBINARES. Tube-nosed Swimmers.

Family PROCELLARIIDÆ. Fulmars and Shearwaters.

Subfamily PROCELLARIINÆ. Fulmars.

Genus PUFFINUS Brisson.

36. (89). Puffinus gravis (*O'Reilly*). Greater Shearwater.

The Shearwaters are birds of the open sea, but as they occur off our coast they can be given a place in the list, although it is doubtful if they ever occur within the three mile limit of the coast. As their occurrence off our coast has been satisfactorily demonstrated, this species and the succeeding one are accordingly given a place in the list.

County Records.—Cumberland, given in Brown's Catalogue of the Birds of Portland as "being said to be rather common by fishermen"; Washington, "common at sea" (Boardman).

37. (94). Puffinus stricklandi *Ridgw.* Sooty Shearwater.

The notes on the preceding species will apply equally well here.

County Records.—Cumberland, "rarely seen except long distances from land, there apparently common" (Brown's Cat. Birds of Portland, p. 35); Washington, "rare at sea" (Boardman).

Genus DAPTION Stephens.

38. (102). Daption capensis (*Linn.*). Pintado Petrel.

A specimen is recorded by Mr. H. A. Purdie in Stearn's "New England Bird Life," p. 387. In view of recent information from Mr. Purdie, the locality given in the above as Harpswell would now seem to be erroneous. In a recent letter under the date of March 2d, he writes: "I happened to be in Worcester yesterday on business. I went to the Natural History Rooms and saw my bird still labeled *Manx Shearwater*, Lewiston, Me., June, 1875, and this was the date and place that a Mr. Dickinson of the Worcester Natural History Society purchased the bird with three or four others of one Levi Wooley, who I understand then called it *Manx Shearwater* and so it has stood ever since. Said Wooley now lives in Waltham, and it seems has lately been to see the bird and swears it is the bird he sold Dickinson in 1875, and I presume still believes it to be as first identified by him. It is not unlikely that the locality named, Harpswell, is wrong. At any rate Mr. Wooley now declares that the bird was shot by Mr. C. F. Nason at Lake Mooseluckmeguntic in September, 1872. Nason, Wooley and two other men were of the party at the lake. Wooley saved and brought it home with him. Some of the other skins that Dickinson got of him may have been shot at Harpswell, so by some misunderstanding the locality of the bird in question got mixed."

From the foregoing it would seem that this bird is undeniably entitled to a place in our list as an accidental visitor.

Genus OCEANODROMA Reichenbach.

39. (106). Oceanodroma leucorhoa (*Vieill.*). Leach's Petrel.

This Petrel is the only breeding representative of its order along our coast. It nests abundantly on many of the extreme outer islands of the coast. Especially large colonies nest annually on Seal, Big Spoon, Little Duck and Green Islands. A burrow from one to three feet in length is excavated in the soft loamy soil of the island selected for a breeding place, and at the end of this hole a small quantity of dry grass, leaves, rootlets or other accessible material is rudely shaped into some resemblance to a nest. Here the parent bird may be found, incubating its single white egg which in most cases is spotted or wreathed with various obscure reddish or lavender markings about the larger end. The eggs are deposited in late June or July, and the parent remains on the nest until removed by hand in case some person digs out a burrow. When taken in the hand they spit forth a quantity of clear, yellowish, musky smelling liquid. When this falls on one's clothes the odor is perceptible for a long time, and the eggs of this species retain the characteristic smell for years. Birds are not often seen in the day time on the islands where they breed, unless one opens the nesting place and forcibly removes the parent. On being turned loose in such cases, the bird seems dazed and stumbles about in a semi-drunken manner or stupidly thrusts its head into the nearest cavity. If tossed into the air it heads straight out to sea. It is noteworthy that in cases where the nest contains its egg only one bird will be found incubating, its mate being far out to sea. In cases where the nest is still in the process of construction, both birds are usually found occupying it. At night the islands where these birds nest become full of manifestations of life. The sitting birds leave their nests and go to feed, while their mates who have passed the day at sea relieve them on the nests. This Petrel is resident off the coast, but in winter is usually found far out at sea. It is often blown inland by severe storms.

County Records.—Androscoggin, "one taken at Lake Auburn and now in the collection of Prof. Stanton" (Stanton in epist.); Cumberland, "breeds on Green Islands, Casco Bay, common" (Brown in Bull. Nutt. Orn. Club, Vol. 2, p. 28); Hancock, "nests on many of the outer islands" (Knight); Knox, "summer" (Rackliff); Oxford, "one shot on Lake Pennesseewassee, Oct. 21st, 1896, by Will Gary" (Oxford Co. Advertiser of that date or thereabouts); Penobscot, "accidental, four specimens have been taken to my knowledge" (Knight); Sagadahoc, "common July and Aug." (Spinney); Washington, "common, breeds on the islands" (Boardman).

Subfamily OCEANITINÆ.

Genus OCEANITES Keyserling and Blasius.

40. (109). Oceanites oceanicus (Kuhl). Wilson's Petrel.

This species is a summer visitor off our coast and is then of uncommon occurrence at sea. It nests on Kerguelen Island, off the coast of Africa, in February, and consequently its proper rating on our list is as a casual visitor.

County Records.—Cumberland, "appears uncommon" (Brown's Cat. Birds of Portland, p. 35); Sagadahoc, "rare" (Spinney); Washington, "rare, seen only in summer" (Boardman).

Order STEGANOPODES. Totipalmate Swimmers.

Family SULIDÆ. Gannets.

Genus SULA Brisson.

Subgenus DYSPORUS Illiger.

41. (117). Sula bassana (Linn.). Gannet.

An irregular migrant and winter resident along the coast, but seemingly never abundant. It nests north of the state, being formerly common on the Gannet Rocks in the Gulf of St. Lawrence.

County Records.—Cumberland, "winter resident, apparently common" (Brown's Cat. of Birds of Portland, p. 33); Kennebec, "accidental, one specimen" (Hamlin's List of Birds of Kennebec Co., p. 173 of the 10th Annual Report Sec'y Me. Bd. Agr.); Knox, "very rare" (Rackliff); Sagadahoc, "common spring and fall" (Spinney); Washington, "common down the bay" (Boardman).

Family PHALACROCORACIDÆ. Cormorants.

Genus PHALACROCORAX Brisson.

Subgenus PHALACROCORAX.

42. (119). Phalacrocorax carbo (*Linn.*). Cormorant.

A common winter visitor along the coast but to my knowledge it certainly does not breed in the state. It departs for the north in late April or early May.

County Records.—Cumberland, "apparently a common winter resident" (Brown's Cat. Birds of Portland, p. 33); Knox, "winter" (Rackliff); Penobscot, "accidental, one shot at Chemo Pond in October, 1896, by Mr. Mudgett of Orono" (Knight); Sagadahoc, "a few from fall to spring" (Spinney); Washington, "not abundant" (Boardman).

43. (120). Phalacrocorax dilophus (*Swain.*). Double-crested Cormorant.

This species is commonest coastwise in migrations, but it is also a rare resident along the coast. A few pairs nest annually on Black Horse Ledge near Isle au Haut. It is, however, seemingly not found in winter, save in these counties near or within the Alleghanian Fauna.

County Records.—Cumberland, "apparently an uncommon winter resident" (Brown's Cat. Birds of Portland, p. 33); Hancock, "rare summer resident on coast, breeds, much commoner as a migrant" (Knight); Knox, "migrant" (Rackliff); Penobscot, "one was shot at Kingman about Nov. 18th, 1895, by Rev. J. W. Hatch, it is now in the University of Maine collection" (Knight); Sagadahoc, "a few from fall to spring" (Spinney); Washington, "common in migration" (Boardman).

Family PELECANIDÆ. Pelicans.

Genus PELECANUS Linnæus.

Subgenus CRYTOPELICANUS Reichenbach.

44. (125). Pelecanus erythrorhynchos *Gmel.* American White Pelican.

Of accidental occurrence in the state where two specimens have been captured.

County Records.—Penobscot, "one shot on Passadumkeag Stream, near Saponic Lake, May 28th, 1892, by Peter Sibley, it is now in my collection (Hardy); Washington, "one seen at Calais, it was afterward shot over the line in New Brunswick" (Boardman).

Order ANSERES.   Lamellirostral Swimmers.

Family ANATIDÆ.   Ducks, Geese, and Swans.

Subfamily MERGINÆ.   Mergansers.

Genus MERGANSER Brisson.

45.   (129.)   Merganser americanus (*Cass*.).   American Merganser.

While this species is to be found in the state throughout the year, it is by no means a resident of one particular locality for this time.   It is a fairly common migrant and winter resident along the coast, while in the interior it is a summer resident and breeder on some of the lakes in the northern part of the state.

County Records.—Androscoggin, "abundant migrant" (Johnson); Cumberland, "common transient" (Mead), "common" (Brock); Franklin, "common summer resident"(Richards); Hancock, "winter"(Knight); Kennebec, "rare" (Dill); Knox, "winter resident" (Racklifl); Oxford, "common migrant" (Johnson); Penobscot, "quite rare migrant" (Knight); Piscataquis, "common, breeds" (Homer); Sagadahoc, "quite plenty from fall to spring" (Spinney); Somerset, "not very common, apparently only migrant" (Morrell); Waldo, "winter resident, seemingly not very common" (Knight); Washington, "not rare, breeds" (Boardman); York, "rare migrant" (Adams).

46.   (130).   Merganser serrator (*Linn*.).   Red-breasted Merganser.

This species is quite a common resident along such parts of the coast as belong to the Canadian Fauna, and also occurs as a summer resident on some of the interior lakes.   In the Alleghanian Fauna it occurs as a migrant, or winter resident.   For notes on the breeding of this species along the coast see Knight, The Auk, Vol. 12, p. 387.

County Records.—Androscoggin, "fairly common migrant" (Johnson); Aroostook, "common, breeding near Houlton" (Batchelder Bull. Nutt. Orn. Club, Vol. 7, p. 152); Cumberland, "common" (Brock); Franklin, "rare migrant" (Richards); Hancock, "breeds quite commonly among the islands, rare in winter" (Knight); Kennebec, (Dill); Knox, "resident" (Racklifl); Oxford, "quite common" (Nash); Penobscot, "nests" (Hardy); Piscataquis, "quite common summer resident" (Whitman); Sagadahoc, "quite plenty from fall to spring" (Spinney); Waldo, "not very common, I do not believe it nests within the county" (Knight); Washington, "not rare" (Boardman); York, (Butters).

Genus LOPHODYTES Reichenbach.

47. (131). Lophodytes cucullatus (*Linn.*). Hooded Merganser.

While this species is quite generally distributed throughout the state in the migrations, it cannot be said to be common anywhere. It breeds on some of our ponds and lakes.

County Records.—Androscoggin, "fairly common migrant" (Johnson); Cumberland, "rare" (Mead), "common" (Brock); Franklin, "rare migrant" (Richards); Hancock, "rare migrant" (Dorr); Kennebec, "very rare" (Dill); Knox, "rare migrant" (Racklift); Oxford, "breeds at Lake Umbagog" (Maynard's List of Birds of Coos Co., N. H., and Oxford Co., Me., p. 30); Penobscot, "rare migrant" (Knight); Piscataquis, "common, breeds" (Homer); Sagadahoc, "very few from fall to spring" (Spinney); Somerset, "rare migrant" (Morrell); Washington, "not rare, breeds" (Boardman).

Subfamily ANATINÆ. River Ducks.

Genus ANAS Linnæus.

48. (132). Anas boschas *Linn.* Mallard.

The Mallard may be classed as a rare migrant throughout the state, and an occasional winter resident along the coast.

County Records.—Androscoggin, "rare migrant" (Johnson); Cumberland, "rare" (Mead), "uncommon, chiefly transient, occasionally occurring in winter" (Brown's Cat. Birds of Portland, p. 30); Hancock, "rare" (Dorr); Kennebec, "very scarce" (Dill); Knox, "rare in winter" (Racklift); Oxford, "very rare" (Nash); Penobscot, "I shot one last year" (Hardy); Sagadahoc, "very few, fall to spring" (Spinney); Somerset, "accidental, one shot Nov. 7th, 1893" (Morrell); Washington, "accidental" (Boardman).

49. (133). Anas obscura *Gmel.* Black Duck.

A very common summer resident on many of our streams and lakes. Along the coast it is resident although much rarer in winter than at other seasons.

County Records—Androscoggin, "common summer resident" (Johnson); Aroostook, "breeds" (Batchelder in Bull. Nut. Orn. Club, Vol. 7, p. 151); Cumberland, "common" (Brock); Franklin, "common summer resident" (Swain); Hancock, "common resident" (Dorr); Kennebec (Dill); Knox, "resident" (Racklift); Oxford, "breeds" (Nash); Penobscot, "breeds commonly along secluded streams and ponds" (Knight); Piscataquis, "common, breeds" (Homer); Sagadahoc, "plenty, a few in summer" (Spinney); Somerset, "common summer resident" (Morrell); Waldo, "breeds to some extent" (Knight); Washington, "common" (Boardman); York, "migrant" (Adams).

Subgenus CHAULELASMUS Bonaparte.

50. (135). Anas strepera *Linn.* Gadwall.

An accidental visitor to the state which has only been recorded from two counties.

County Records.—Cumberland, "two specimens, April 29, 1879" (Smith's List of Birds of Maine, Forest and Stream, Vol. 20, p. 125); Washington, "accidental" (Boardman).

Subgenus MARECA Stephens.

51. (137). Anas americana *Gmel.* Baldpate.

A not uncommon migrant along the coast, rarer in the interior. Breeds north of our limits.

County Records.—Androscoggin, "migrant" (Johnson); Cumberland, "often common" (Brock); Hancock, "rare" (Dorr); Oxford, "very rare" (Nash); Penobscot, "one shot at Monument Brook" (Hardy); Sagadahoc, "very few spring and fall" (Spinney); Washington "very rare" (Boardman).

Subgenus NETTION Kaup.

52. (139). Anas carolinensis *Gmel.* Green-winged Teal.

A quite common migrant throughout the state, occurs in greater numbers in the fall. It is not known to nest in the state.

County Records.—Androscoggin, "fairly common migrant" (Johnson); Cumberland, "common" (Brock); Hancock, "rare" (Dorr); Kennebec (Dill); Knox, "rare migrant" (Rackliff); Oxford, "common" (Nash); Penobscot, "rare in migrations" (Lord); Sagadahoc, "very few spring and fall" (Spinney); Somerset, "not common migrant" (Morrell); Washington, "not common" (Boardman).

Subgenus QUERQUEDULA Stephens.

53. (140). Anas discors *Linn.* Blue-winged Teal.

This little Duck occurs chiefly as a migrant, and is quite common in the spring and fall. A few remain to nest in the extreme northern and eastern counties.

County Records.—Androscoggin, "fairly common migrant" (Johnson); Cumberland, "common" (Brock); Franklin, "rare migrant" (Richards); Hancock, (Dorr); Kennebec, (Dill); Knox, "rare migrant" (Rackliff); Oxford, "quite common" (Nash); Penobscot, "seemingly a quite rare migrant" (Knight); Piscataquis, "rare" (Homer); Sagadahoc, "common spring and fall" (Spinney); Somerset, "common migrant" (Morrell); Washington, "common, breeds" (Boardman).

Genus SPATULA Boie.

54. (142). Spatula clypeata (*Linn.*). Shoveller.

A rare migrant along our coast; most of the specimens recorded seem to be from Cumberland County.

County Records.—Cumberland, "rare" (Brock); "six specimens are recorded from this county" (for these records see Smith's List of Birds of Maine in Forest and Stream, Vol. 20, p. 125); Sagadahoc, "rare spring and fall" (Spinney); Washington, "accidental" (Boardman).

Genus DAFILA Stephens.

55. (143). Dafila acuta (*Linn.*). Pintail.

A rare migrant throughout the state, but somewhat commoner along the coast than in the interior. So infrequent is this species in occurrence in some parts of the state that it has been recorded recorded as accidental by one of our best observers.

County Records.—Androscoggin, "rare migrant" (Johnson); Cumber land, "fairly common" (Brock); Franklin, "rare migrant" (Richards); Kennebec, (Dill); Knox, "rare migrant" (Rackliff); Oxford, "very rare" (Nash); Sagadahoc, "rare spring and fall" (Spinney); Somerset, "rare, two specimens in fall of '95" (Morrell); Washington, "accidental" (Boardman).

Genus AIX Boie.

56. (144). Aix sponsa (*Linn.*). Wood Duck.

A common summer resident in unsettled localities along streams, ponds, and lakes throughout the state. It is now less common than of former years.

County Records.—Androscoggin, "summer resident" (Johnson); Aroostook, "breeds" (Batchelder in Bull. Nutt. Orn. Club, Vol. 7, p. 151); Cumberland, "common" (Mead); "rather common transient, a few remain through summer" (Brown's Cat. Birds of Portland, p. 31); Franklin, "rare summer resident" (Richards); Hancock, "summer resi- dent" (Murch); Kennebec, "common" (Dill); Knox, "summer" (Rack- liff); Oxford, "breeds commonly" (Nash); Penobscot, "summer resi- dent but not so common as in former years" (Knight); Piscataquis, "common, breeds" (Homer); Sagadahoc, "quite common in fall" (Spinney); Somerset, "rare summer resident, common migrant" (Mor- rell); Washington, "common" (Boardman); York, "migrant, possibly a few breed" (Adams).

Subfamily FULIGULINÆ.   Sea Ducks.

Genus AYTHYA Boie.

57.   (146).   Aythya americana (*Eyt.*).   Redhead.

A rare migrant through the greater part of the state.   It is reported as breeding in Washington County.

County Records.—Cumberland, "fairly common" (Brock); Kennebec, (Dill); Knox, "rare migrant" (Racklift); Oxford, "not very common" (Nash); Penobscot, "one shot at Levant, October 26, 1896, and now in the collection of the University of Maine" (Knight); Washington, "rare, breeds" (Boardman).

58.   (147).   Aythya vallisneria (*Wils.*).   Canvas-back.

As there are but two seemingly authentic records for the state, this species may be classed as a casual visitor without much doubt. There have been other specimens reported, but on investigation they have proved to be the preceding species.   The specimen of Redhead recorded from Penobscot County was at first reported as a Canvas-back, but after investigation was found not to be this species.   A number of cases reported, where the report was not substantiated by the production of the specimen for identification, have therefore been deemed not worthy of recording.

County Records.—Cumberland, "taken in Casco Bay and at Cape Elizabeth" (Smith's List of Birds of Maine, Forest and Stream, Vol. 20, p. 184).

Subgenus FULIGULA Stephens.

59.   (148).   Aythya marila nearctica *Stejn*.   American Scaup Duck.

Common coastwise in migrations, and also occurs some winters. Usually found in large flocks.

County Records.—Androscoggin, (Johnson); Cumberland, "common" (Brock); Hancock, "a specimen from this county in my collection" (Knight); Knox, "winter" (Racklift); Sagadahoc, "quite common in fall" (Spinney); Washington, "not common" (Boardman).

60.   (149).   Aythya affinis (*Eyt.*).   Lesser Scaup Duck.

A rare migrant along the coast, wintering chiefly in the south and breeding north of the state.   From its close resemblance to the preceding species it is likely to escape observation except by persons well acquainted with the species.

County Records.—Cumberland, "rare" (Brock); Knox, "rare in winter" (Rackliff); Washington, "not common" (Boardman).

61. (150). Aythya collaris (*Donov.*). Ring-necked Duck.

A rare migrant of somewhat local occurrence along the coast. It is reported as breeding in Washington County.

County Records.—Cumberland, "rare, has occurred between March 31st and May 1st" (Brown's Cat. Birds of Portland, p. 32); Washington, "not common, breeds" (Boardman).

Genus CLANGULA Leach.

62. (151). Clangula clangula americana (*Bonap.*). American Golden-eye.

Of quite general distribution in the migrations, and also found throughout the winter along the coast. It breeds on some of the ponds and lakes of the northern part of the state.

County Records.—Androscoggin, "common migrant" (Johnson); Cumberland, "common" (Brock); Hancock, "common in winter" (Dorr); Kennebec (Dill); Knox, "winter" (Rackliff); Oxford, "breeds" (Maynard's List of Birds of Coos Co., N. H., and Oxford Co., Me., p. 29); Penobscot, "common migrant, have seen it on the ponds as late as the middle of May so it may possibly nest in the county" (Knight); Piscataquis, "common, breeds" (Homer); Sagadahoc, "quite common in winter" (Spinney); Somerset, "common migrant, a male seen June 23, 1896" (Morrell); Washington, "common resident" (Boardman).

63. (152). Clangula islandica (*Gmel.*). Barrow's Golden-eye.

A rare winter visitor to our coast, but on account of its general resemblance to the preceding it would be likely to escape detection except by persons well acquainted with our birds.

County Records.—Hancock, "rare winter resident" (Dorr); Knox, "winter" (Norton); Washington, "common in winter" (Boardman).

Genus CHARITONETTA Stejneger.

64. (153). Charitonetta albeola (*Linn.*). Buffle-head.

A very common migrant in most parts of the state. It winters along the coast in the southern part of the state, and is reported as breeding in the northeastern part.

County Records.—Androscoggin, "common migrant" (Johnson); Cumberland, "rare" (Mead), "common" (Brock); Franklin, "common migrant" (Richards); Knox, "winter" (Rackliff); Oxford, "common visitant" (Nash); Penobscot, "have seen specimens taken in the county" (Knight); Piscataquis, "common migrant" (Homer); Sagadahoc, "formerly plenty, now scarce" (Spinney); Washington, "common spring and fall, breeds" (Boardman).

Genus HARELDA Stephens.

65. (154). Harelda hyemalis (*Linn.*). Old-squaw.

Very common along the coast in autumn, winter and spring; also of rare occurrence in summer, but does not breed within the state. Birds seen in summer are probably crippled or barren individuals which have not accompanied their relatives to the breeding grounds in the far north.

County Records.—Androscoggin, "migrant" (Johnson); Cumberland, "normally a winter resident, many individuals supposed to be crippled remain all summer" (Brown's Cat. Birds of Portland, p. 32); Hancock, "common in winter" (Dorr); Knox, "winter" (Rackliff); Oxford, "rare visitant" (Nash); Penobscot, "one shot at East Eddington" (Hardy); Piscataquis, "rare migrant" (Homer); Sagadahoc, "common in winter, have also seen it here in summer" (Spinney); Washington, "abundant" (Boardman).

Genus HISTRIONICUS Lesson.

66. (155). Histrionicus histrionicus (*Linn.*). Harlequin Duck.

Formerly quite a common winter resident along the coast, but now occurs in greatly diminished numbers. At present it is to be found in numbers only on the eastern half of the coast, and even here it is not very common. The habits and occurrence of this species along our coast form the subject of an admirable article by Mr. A. H. Norton (Cf. Norton, The Auk, Vol. 13, pp. 229-234).

County Records.—Cumberland, "rare winter visitant" (Brown's Cat. Birds of Portland, p. 32); Hancock, "the 'Lord and Lady Ducks' are reported by gunners as fairly common in winter in the vicinity of various small islands near Isle au Haut" (Knight); Knox, "winter" (Rackliff); Sagadahoc, "rare in winter, formerly plenty" (Spinney); Washington, "among the islands in fall and winter" (Boardman).

Genus SOMATERIA Leach.

Subgenus SOMATERIA.

67. (159). Somateria mollissima borealis *A. E. Brehm*. Northern Eider.

This is a rare winter visitor along our coast, but still it is doubtless much commoner than would be supposed from the meager
records presented. Breeds in the far north, and south as far as
Labrador.

County Records.—Cumberland (Brock); Knox, "rare in winter"
(Rackliff); Sagadahoc, "one specimen, a male" (Spinney).

68. (160). Somateria dresseri *Sharpe*. American Eider.

Of general occurrence as a winter visitor along the entire coast.
It is a rare summer resident from Isle au Haut eastwŕrd, breeding
on some of the smaller islands in colonies of two to four pairs of
birds or by single pairs. It used to be much commoner than it is
now, and its ultimate extinction as a breeding bird is only a question of a few years. The fishermen know of every "Sea Duck's"
(this is the name they have for the bird) nest in the neighborhood,
and promptly rob the nests of their eggs for culinary use, or to
hatch and rear them with their domestic fowls. (Cf. Knight, The
Auk, Vol. 12, p. 388).

County Records.—Cumberland, "fairly common" (Brock); Hancock,
"in 1896 about seven pair of these birds nested on various small islands
between Little Duck and Isle au Haut, while the previous year at least
ten pair of the birds were found in the same locality; their numbers are
decreasing yearly" (Knight); Knox, "resident" (Rackliff); Sagadahoc,
"common in winter, arrives in November" (Spinney); Washington,
"abundant in winter" (Boardman).

Subgenus ERIONETTA Coues.

69. (162). Somateria spectabilis (*Linn.*). King Eider.

A regular winter resident along the coast, where it sometimes
occurs quite commonly, but is usually rare. It breeds in the
Arctic regions.

County Records.—Cumberland, "fairly common" (Brock); Knox,
"winter" (Rackliff); Sagadahoc, "three specimens in ten years") Spinney); Washington, "not rare in winter" (Boardman).

Genus OIDEMIA Fleming.
Subgenus OIDEMIA.

70. (163). Oidemia americana Sw. and Rich. American Scoter.

A resident along the coast, but commonest in the migrations and
in winter. It is of rare occurrence inland on the ponds and lakes.

The birds seen in summer have probably been wounded so as to be unable to migrate northward with their relatives, or are possibly barren individuals who have no desire to do so. At any events birds taken in the summer, which seem perfectly strong and able to fly well, have on dissection shown no indications of breeding. It certainly has never been detected breeding within the state, and in Labrador and other northern countries where it is known to breed, it is said to resort to the ponds and streams away from the seacoast for this purpose.

County Records.—Androscoggin, "fairly common migrant" (Johnson); Cumberland, "common" (Brock); Hancock, "common in winter, rare in summer" (Knight); Knox, "resident" (Rackliff); Penobscot (Hardy); Sagadahoc, "seen the year around" (Spinney); Waldo, (Knight); Washington, "common" (Boardman); York, (Butters).

Subgenus MELANITTA Boie.

71. (165). Oidemia deglandi *Bonap.* White-winged Scoter.

The remarks made under the preceding species will all apply equally well here. It does not breed in the state.

County Records.—Androscoggin, "rare migrant" (Johnson); Cumberland, "common" (Brock); Hancock, "common in fall, winter, and spring, fairly common in summer" (Knight); Knox, "resident" (Rackliff); Penobscot, "a small flock seen near the mouth of Pushaw Pond late in April, 1896" (Knight); Sagadahoc, "seen the year around" (Spinney); Waldo, (Knight); Washington, "common" (Boardman).

Subgenus PELIONETTA Kaup.

72. (166). Oidemia perspicillata (*Linn.*). Surf Scoter.

Occurs under the same conditions as the American Scoter does, and likewise does not nest in the state. Seemingly not so common in summer as the other two species, or at least this is the case in Hancock County.

County Records. — Androscoggin, (Pike); Cumberland, "common" (Brock); Franklin, "rare migrant" (Richards); Hancock, "common in winter" (Knight); Knox, "resident" (Rackliff); Oxford, "visitant" (Nash); Penobscot, (Hardy); Piscataquis, "rare" (Homer); Sagadahoc, "seen the year around" (Spinney); Somerset, "flock of nine seen in September 1895, one shot by H. H. Johnson" (Morrell); Waldo, (Knight); Washington, "common" (Boardman).

Genus ERISMATURA Bonaparte.

73. (167). Erismatura jamaicensis (*Gmel.*). Ruddy Duck.

Chiefly occurs as a migrant throughout the state and is never especially common. It breeds in limited numbers in the northeastern part of the state.

County Records.—Androscoggin, "migrant" (Johuson); Cumberland, "common" (Brock); Hancock, "rare" (Dorr); Knox, "transient visitor" (Norton); Penobscot, "occasional" (Hardy); Sagadahoc, "few fall and spring" (Spinney); Washington, "not rare, breeds" (Boardman); York, "I have a specimen from the town of York" (Norton).

Subfamily ANSERINÆ. Geese.

Genus CHEN Boie.

74. (169). Chen hyperborea (Pall.). Lesser Snow Goose.

Of rare occurrence in migrations when they are quite likely to be taken in almost any part of the state where there are fair-sized bodies of water.

County Records.—Cumberland, "taken in December, 1880" (Brown's Cat. Birds of Portland, p. 30); Hancock, "very rare, one taken at Toddy Pond, October 4th, 1893, which I sold to Mr. Brewster" (Dorr); Kennebec, "taken at Hallowell on November 25th, 1881" (Smith's List of the Birds of Maine, Forest and Stream, Vol. 20, p. 125); Oxford, in the Auk for April, 1897, p. 207, Mr. Brewster records a specimen taken at Lake Umbagog, Maine, October 2, 1896, by Mr. Charles Douglass; Penobscot, "have one shot at Pushaw and saw one shot at Nicatous" (Hardy).

75. (169a). Chen hyperborea nivalis (*Forst.*). Greater Snow Goose.

An accidental visitor in the migrations. I have been able to find only one authentic New England record of the species. In view of this I will quote from a recent letter from the owner of the specimen, Mr. Chas. F. Batchelder, of Cambridge, Mass. He says: "In reply to yours of the 14th inst., in regard to my note in The Auk (Vol. 7, p. 284), entitled The Snow Goose (*Chen hyperborea nivalis*) on the Coast of Maine, I will say that I there recorded a specimen of this bird which I received in the flesh and which was shot on Heron Island at the mouth of the Kennebec River, April 7th, 1890. It was a female and had been seen about here for three days before it was shot. . . . It is now in my collection."

Genus BRANTA Scopoli.

76. (172). Branta canadensis (*Linn.*). Canada Goose.

Generally distributed as a migrant throughout the state, and the v-shaped flocks of these birds, led by some old gander, are a common sight in fall and spring. They breed in the north.

County Records.—Androscoggin, "common migrant" (Johnson); Cumberland, "migrant" (Mead); Franklin, "common migrant" (Richards); Hancock, "migrant" (Murch); Kennebec, "migrant" (Gardiner Branch); Knox, "migrant" (Rackliff); Oxford, "common migrant" (Johnson); Penobscot, "flocks flying over head are a common sight in the spring and fall migrations" (Knight); Piscataquis, "migrant" (Homer); Sagadahoc, "common in migration" (Spinney); Somerset, "not common migrant" (Morrell); Waldo, "migrant" (Spratt); Washington, "common" (Boardman); York, "rare migrant" (Adams).

77. (172 a). Branta canadensis hutchinsii (Rich.). Hutchin's Goose.

While the evidence at hand will allow us to rank the species as accidental only it is highly probable that these birds may occur quite regularly in the migrations. This species is cited by Smith as having been shot in Maine. (Cf. Smith, Forest and Stream, Vol. 20, p. 125). Fortunately I have later and more positive evidence of its occurrence, for in a recent letter Dr. Brock of Portland writes: "I can give you the most positive information regarding the Hutchin's Goose, as I have the specimen myself. It is an adult male and was taken at Cape Elizabeth, November 13, 1894. I have seen a specimen said to have been shot in the Rangeley region."

78. (173). Branta bernicla (*Linn.*). Brant.

A common migrant along the coast and of rare occurrence in the interior.

County Records.—Cumberland, "rare" (Brock); Hancock, "spring migrant" (Dorr); Kennebec (given in Hamlin's List of Birds of Waterville, Rep. Sec'y Me. Bd. Agr., 1865, p. 172); Knox, "migrant" (Rackliff); Oxford (Maynard's List of Birds of Coos Co., N. H. and Oxford Co., Me., p. 29); Sagadahoc, "common in migrations" (Spinney); Washington, "common" (Boardman).

Order HERODIONES.   Herons, Storks, Ibises, etc.
  Suborder CICONIÆ.   Storks, etc.
    Family CICONIIDÆ.   Storks and Wood Ibises.
      Subfamily TANTALINÆ.   Wood Ibises.
        Genus TANTALUS Linnæus.
78. 1   (188).   Tantalus loculator *Linn.*   Wood Ibis.

But one specimen is known to have been taken in New England. This was shot in Berwick, York County, Maine, July 16th, 1896, by H. M. Brackett, and is now in the collection of Prof. J. Y. Stanton of Lewiston.

  Suborder HERODII.   Herons, Egrets, Bitterns, etc.
    Family ARDEIDÆ.   Herons, Bitterns, etc.
      Subfamily BOTAURINÆ.   Bitterns.
        Genus BOTAURUS Hermann.
79.   (190).   Botaurus lentiginosus (*Montag.*).   American Bittern.

A common summer resident, breeds in the vicinity of ponds, meadows along streams, and cat-tail swamps, throughout the state. It is known to the average individual under the elegant names of Shite-poke, Stake Driver, Bog Hen, Thunder Pump and Indian Hen. Usually only one pair of birds will be found in a given locality, while perhaps half or quarter of a mile away another pair will likewise lay claim to that territory.

County Records.—Androscoggin, "common summer resident" (Johnson); Aroostook, "common" (Batchelder in Bull. Nutt. Orn. Club, Vol. 7, p. 151); Cumberland, "common summer resident" (Mead); Franklin, "common summer resident" (Lee & McLain); Hancock, "summer resident" (Murch); Kennebec, (given in Hamlin's List of the Birds of Waterville, Rep. Sec'y Me. Bd. Agr., 1865, p. 172); Knox, "summer" (Rackliff(; Oxford, "breeds commonly" (Nash); Penobscot, "breeds in suitable localities throughout the county" (Knight); Piscataquis, "common, breeds" (Homer); Sagadahoc, "common" (Spinney); Somerset, "common summer resident" (Morrell); Waldo, (Spratt); Washington, "very common in summer" (Boardman); York, "breeds" (Adams).

Genus ARDETTA Gray.
80.   (191).   Ardetta exilis (*Gmel.*).   Least Bittern.

Formerly quite a common summer resident of those parts of the state belonging to the Alleghanian fauna, but of late years it has been rare.

County Records.—Androscoggin, "rare summer resident" (Johnson); Cumberland, "found breeding at Falmouth in 1863" (Smith's List of the Birds of Maine, Forest and Stream, Vol. 20, p. 105); Knox, "rare migrant" (Racklift); Sagadahoc, "rare, two specimens" (Spinney); Washington, "rare" (Boardman).

Subfamily ARDEINÆ. Herons and Egrets.
Genus ARDEA Linnæus.
Subgenus ARDEA.

81. (194). Ardea herodias *Linn.* Great Blue Heron.

A very common summer resident throughout the state. Colonies of from three to one hundred pairs of birds may still be found nesting in isolated localities about ponds and lakes. In such places they seem to prefer to place their nests in tall rock maple or beech trees, and as many as six nests are often seen in one tree. Many small colonies also breed on the numerous wooded islands along the seacoast, and here the nests are almost invariably placed in spruce trees, or at least this is the case among the islands of Penobscot Bay.

County Records.—Androscoggin, "fairly common summer resident" (Johnson); Aroostook, "common" (Batchelder, Bull. Nutt. Orn. Club, Vol. 7. p. 151); Cumberland, "common summer resident" (Mead); Franklin, "common summer resident" (Swain); Hancock, "breeds commonly on three or four islands of Penobscot Bay and about the ponds of the interior" (Knight); Kennebec, "common summer resident" (Gardiner Branch); Knox, "summer" (Racklift); Oxford, "breeds commonly' (Nash); Penobscot, "I know of at least two heroneries occupied yearly, and perhaps there are fifty or sixty pair of birds at the two places" (Knight); Piscataquis, "common, breeds" (Homer); Sagadahoc, "common fall and spring" (Spinney); Somerset, "common summer resident" (Morrell); Waldo, "summer resident" (Knight); Washington, "common" (Boardman); York, (Adams).

Subgenus HERODIAS Boie.

82. (196). Ardea egretta *Gmel.* American Egret.

An accidental visitor from the south; four specimens have been taken within the state, while Mr. Boardman has also taken one at Grand Menan.

County Records.—Cumberland, "one taken August 22nd, 1853" (Smith's List of the Birds of Maine, Forest and Stream, Vol. 20, p. 104), "one at Scarborough in April, 1875" (Rod and Gun, Vol. 6, p. 65); Hancock, "one shot at Cranberry Island on April 7th, 1891, by Elwood Richardson" (E. Smith); Kennebec, "one shot by Will Libby on Pleasant Pond, between Richmond and West Gardiner, August 20th, 1896, this was in company with another of the same species" (Powers).

Subgenus Florida Baird.

83. (200).   Ardea cærulea *Linn*.   Little Blue Heron.

An accidental visitor from the south, of which only one speci-
men is recorded.   This was taken at Scarborough, Cumberland
County, September, 1881, and is recorded by Brown in the Bull.
Nutt. Orn. Club, Vol. 7, p. 123.

Subgenus BUTORIDES Blyth.

84. (201).   Ardea virescens *Linn*.   Green Heron.

A rare summer resident in the southwestern part of the state,
and of accidental occurrence in the eastern part.

County Records.—Androscoggin, "fairly common migrant" (Johnson);
Cumberland, "uncommon summer resident near Portland" (Brown's Cat.
Birds of Portland, p. 24); Franklin, "rare" (Richards); Kennebec, "very
rare" (Robbins); Knox, "summer" (Rackliff); Oxford, "very rare"
(Nash); Penobscot, "have seen one taken here" (Hardy); Sagadahoc,
"common summer resident, breeds" (Spinney); Somerset, "quite common, apparently only migrant" (Morrell); Washington, "rare" (Boardman).

Genus NYCTICORAX Stephens.

Subgenus NYCTICORAX.

**85. (202).   Nycticorax nycticorax nævius (*Bodd.*).   Black-
crowned Night Heron.**

A common summer resident along the coast, where it breeds in
colonies on some of the wooded islands.   Not so common in the
interior about the ponds and lakes where it probably also nests.

County Records.—Androscoggin, "fairly common summer resident"
(Johnson); Aroostook, "not common at Houlton" (Batchelder, Bull.
Nutt. Orn. Club, Vol. 7, p. 151); Cumberland, "common" (Brock);
Franklin, "accidental" (Swain); Hancock, "know of two fairly large
colonies nesting on islands in Penobscot Bay" (Knight); Kennebec
(given in Hamlin's List of the Birds of Waterville, Rep. Sec'y Me.
Bd. Agr., 1865, p. 172); Knox, "summer" (Rackliff); Lincoln, "1895,
one" (Norton); Oxford, "rare visitant" (Nash); Penobscot, "plenty"
(Hardy); Piscataquis, "not an uncommon visitor" (Homer); Sagadahoc,
"summer resident" (Spinney); Somerset, "accidental, two specimens
on Aug. 9th, 1896" (Morrell); Waldo, (Knight); Washington, "rare"
(Boardman).

Order PALUDICOLÆ.   Cranes, Rails, etc.

 Suborder RALLI.   Rails, Gallinules, Coots, etc.

  Family RALLIDÆ.   Rails, Gallinules, and Coots.

   Subfamily RALLINÆ.   Rails.

    Genus RALLUS Linnæus.

86. (208). Rallus elegans *Aud.* King Rail.

An accidental visitor to the state.   Only two specimens recorded up to date.

County Records.—Cumberland, "taken at Scarborough" (Brown in Bull. Nutt. Orn. Club, Vol. 7, p. 60), "one shot at Falmouth by Walter Rich, Sept. 19th, 1895" (Brock in the Auk, Vol. 13, p. 79).

87. (211). Rallus crepitans *Gmel.* Clapper Rail.

An accidental visitor from the south which has only been taken in the extreme southern part of the state.

County Records.—Androscoggin, "one taken at Sabattus Pond, in 1874, by C. F. Nasou" (Smith's List of Birds of Maine, Forest and Stream, Vol. 20, p. 124); Cumberland, "rare" (Brown's Cat. Birds of Portland, p. 30); York, (Brown in Bull. Nutt. Orn. Club, Vol. 4, p. 108).

88. (212). Rallus virginianus *Linn.* Virginia Rail.

A rare summer resident in most parts of the state, and probably breeds wherever found in the summer.   It frequents meadows and marshes near the various ponds, and sluggish streams running therefrom.

County Records.—Androscoggin, "rare summer resident" (Johnson); Cumberland, "rare summer resident" (Mead); Franklin, "rare summer resident" (Swain); Kennebec, "W. R. Guilford has a specimen shot at Waterville" (Morrell); Knox, "rare migrant" (Rackliff); Oxford, "found young in down at Fryeburg in August, 1883" (Mead); Penobscot, "rare" (Knight); Sagadahoc, "common in fall" (Spinney); Somerset, "found a nest with five newly hatched young and two sterile eggs at Hartland on August 5, 1896" (Knight); Washington, "common" (Boardman).

    Genus PORZANA Vieillot.

     Subgenus PORZANA.

89. (214). Porzana carolina (*Linn.*). Sora.

This species is of quite general distribution throughout the state, and while of rare or local occurrence in some counties it is quite common in others.

County Records.—Androscoggin, "rare summer resident" (Johnson);
Aroostook, "seen at Fort Fairfield" (Batchelder, Bull. Nutt. Orn. Club,
Vol. 7, p. 151); Cumberland, "common" (Brock); Knox, "migrant"
(Rackliff); Oxford, "not common summer resident at Norway" (Verrill's Birds of Norway, in the Proceedings of the Essex Institute, Vol. 3,
p. 136 et seq.); Penobscot, "it has been reported to me as occurring, by
good authorities, but I cannot say from my own experience that it actually
does" (Knight); Somerset, "common summer resident" (Morrell); Washington, "abundant" (Boardman).

Subgenus COTURNICOPS Bonaparte.

90. (215). Porzana noveboracensis (*Gmel.*). Yellow Rail.

This is probably of rare occurrence throughout the state,
although up to the present it has only been recorded from near the
coast. Owing to the very secretive habits of this bird it might be
very common in a given locality and still not be observed. Mr.
Boardman has found its nest and eggs near Calais.

County Records.—Cumberland, "quite common" (Brock); Knox, "very
rare migrant" (Rackliff); Washington, "rare, breeds" (Boardman).

Genus CREX Bechstein.

91. (217). Crex crex (*Linn.*). Corn Crake.

An accidental visitor from Europe One specimen has been
taken at Dyke Marsh near Falmouth, Cumberland County, October
4, 1889, by John Whitney, and is recorded by Dr Brock in The
Auk, Vol. 13, p. 173.

Subfamily GALLINULINÆ. Gallinules.

Genus IONORNIS Reichenbach.

92. (218). Ionornis martinica (*Linn.*). Purple Gallinule.

An accidental visitor from the south, only four specimens
recorded.

County Records.—Androscoggin," one taken Apr. 11, 1897, at South
Lewiston, by John Turner" (C. D. Farrar); Knox, (Rackliff); Lincoln,
"one taken at Boothbay" (Purdie in Bull. Nutt. Orn. Club, Vol. 5, p.
173); Washington, "accidental" (Boardman).

Genus GALLINULA Brisson.

93. (219). Gallinula galeata (*Licht.*). Florida Gallinule.

A rare migrant in this state. It might almost be classed as a straggler were it not for the fact that there are several records from different parts of the state at different times.

County Records.—Androscoggin, "rare migrant" (Johnson); Cumberland, "taken at Falmouth" (Brock in The Auk, Vol. 13, p. 255); Hancock, "one taken at East Sullivan, May 5th, 1883, by M. Uran" (record from E. Smith); Penobscot, (Hardy); Washington, "several" (Boardman).

Subfamily FULICINÆ. Coots.

Genus FULICA Linnæus.

94. (221). Fulica americana *Gmel.* American Coot.

A somewhat rare migrant in most parts of the state, where it is seemingly noticed oftenest in the autumn. It does not breed in the state, which seems somewhat odd, as this species nests commonly from southern California to Illinois, and parts of Canada, thus not being the inhabitant of any definite faunal area.

County Records.—Androscoggin, "migrant" (Johnson); Cumberland, "rare" (Mead); Hancock, "migrant" (Murch); Kennebec, (Dill); Knox, "rare migrant" (Racklift); Oxford, "visitant" (Nash); Penobscot, "seemingly not rare in fall, have seen a number of specimens shot here" (Knight); Sagadahoc, "common in fall" (Spinney); Somerset, "rare migrant" (Morrell); Washington, "not uncommon" (Boardman).

Order LIMICOLÆ. Shore Birds.

Family PHALAROPODIDÆ. Phalaropes.

Genus CRYMOPHILUS Vieillot.

95. (222). Crymophilus fulicarius (*Linn.*). Red Phalarope.

An uncommon migrant along the coast and of still rarer occurrence in the interior. Mr. Boardman writes: "I have twice found it breeding here," meaning near Calais.

County Records.—Cumberland, "rare, only in migration" (Brock); Penobscot, "I know of a pair being taken at Hermon Pond" (Hardy): Somerset, "accidental, one shot Oct. 17th, 1893" (Morrell); Washington, "not uncommon, a few are summer resident" (Boardman).

Genus PHALAROPUS Brisson.
Subgenus PHALAROPUS.

96. (223). Phalaropus lobatus (*Linn.*). Northern Phalarope.

Quite common as a migrant along the coast and of accidental occurrence in the interior. It nests north of our limits.

County Records.—Cumberland, "rare" (Brock); Franklin, "accidental" (Richards); Hancock, (Dorr); Knox, "migrant" (Norton); Penobscot, "one taken by Mr. Fuller of Newport" (Hardy); Piscataquis, "one shot on the Sebec River, near Milo, May 3, 1897, and sent me in the flesh by William Cooper" (Knight); Sagadahoc, "common" (Spinney); Washington, "plenty spring and fall" (Boardman).

Genus STEGANOPUS Vieillot.

97. (224). Steganopus tricolor *Vieill.* Wilson's Phalarope.

An accidental visitor to this state.

County Records.—Cumberland, "rare, three taken near Scarborough, on June 9th, 1891" (Smith's List of the Birds of Maine, Forest and Stream, Vol. 20, p. 124).

Family RECURVIROSTRIDÆ. Avocets and Stilts.
Genus RECURVIROSTRA Linnæus.

98. (225). Recurvirostra americana *Gmel.* American Avocet.

An accidental visitor from the west of which there is one record. Mr. Boardman has taken it near Calais but in New Brunswick.

County Records.—Cumberland, "one killed on Cape Elizabeth, November 5th, 1878" (Brown, Bull. Nutt. Orn. Club, Vol. 4, p. 108).

Family SCOLOPACIDÆ. Snipes, Sandpipers, etc.
Genus PHILOHELA Gray.

99. (228). Philohela minor (*Gmel.*). American Woodcock.

A common summer resident throughout the state. It arrives in late March or early April. In the fall individuals are occasionally seen as late as the middle of November.

County Records. Androscoggin, "common summer resident" (Johnson); Aroostook, "seen at Fort Fairfield and Houlton" Batchelder, Bull. Nutt. Orn. Club, Vol. 7, p. 151); Cumberland, "common summer resident" (Mead); Franklin, "common summer resident" (Swain); Hancock, "summer resident" (Murch); Kennebec, "summer resident" (Gardiner Branch); Knox, "summer" (Racklift); Oxford, "breeds commonly" (Nash); Penobscot, "breeds commonly in suitable localities" (Knight); Piscataquis, "common, breeds" (Homer); Sagadahoc, "common summer resident" (Spinney); Somerset, "quite common migrant, rare summer resident" (Morrell); Waldo, (Spratt); Washington, "plenty, breeds early" (Boardman); York, "rare breeder" (Adams).

Genus GALLINAGO Leach.

100. (230). Gallinago delicata (*Ord*). Wilson's Snipe.

A common migrant in fall and spring, and a rare summer resident in some parts of the state.

County Records.—Androscoggin, "common migrant" (Johnson); Cumberland, "rare near Bridgton" (Mead), "common" (Brock); Franklin, "common migrant" (Richards); Hancock, "migrant, quite common on the salt marshes" (Knight); Kennebec, "common" (Gardiner Branch); Knox, "migrant" (Racklifl); Oxford, "breeds rarely, found a nest in June, 1881, near Cold River" (Nash); Penobscot, "quite a common migrant, have seen birds along Pushaw Stream in June so it is a rare summer resident" (Knight); Piscataquis, "common migrant" (Homer); Sagadahoc, "rare in spring, common in fall" (Spinney); Somerset, "common migrant, rare summer resident" (Morrell); Washington, "plenty, some breed" (Boardman).

Genus MACRORHAMPHUS Leach.

101. (231). Macrorhamphus griseus (*Gmel.*). Dowitcher.

Quite common along the coast in migrations. They breed in the far north, and begin the migration southward so as to appear on our coast in late July or early August.

County Records.—Cumberland, "common" (Brock); Knox, "summer" (Racklifl); Washington, "rare" (Boardman).

Genus MICROPALAMA Baird.

102. (233). Micropalama himantopus (*Bonap.*). Stilt Sandpiper.

A rare migrant along the coast, and seemingly occurs only in autumn.

County Records.—Cumberland, "transient in autumn only, rather uncommon" (Brown's Cat. Birds of Portland, p 26); Washington, "rare" (Boardman).

Genus TRINGA Linnæus.
Subgenus TRINGA.

103. (234). Tringa canutus *Linn*. Knot.

A quite common migrant coastwise; in the fall migrations it appears in early August. Nests in Arctic regions.

County Records.—Cumberland, "common" (Brock); Hancock, "found it common at Saddleback Ledge, Aug. 19th, 1896" (Knight); Knox, "migrant" (Norton); Sagadahoc, "very rare" (Spinney); Washington, "rare" (Boardman).

104. (235). Tringa maritima *Brünn.* Purple Sandpiper.

A quite common late fall and winter resident among the islands of the coast.

County Records.—Cumberland, "fairly common in winter" (Brock); Knox, "winter" (Rackliff); Sagadahoc, "common in winter" (Spinney); Washington, "abundant in winter" (Boardman).

105. (239). Tringa maculata *Vieill.* Pectoral Sandpiper.

A common fall migrant in many parts of the state, but of rare occurrence in the spring.

County Records.—Cumberland, "common" (Brock); Knox, (Rackliff); Oxford, "not common at Norway in autumn" (Verrill's List of the Birds of Norway); Piscataquis, "rare migrant" (Homer); Sagadahoc. "common in fall" (Spinney); Somerset, "quite common migrant" (Morrell); Washington, "common in fall" (Boardman).

106. (240). Tringa fuscicollis *Vieill.* White-rumped Sandpiper.

A quite rare migrant along the coast and of casual occurrence in the interior.

County Records.—Cumberland, "occasional" (Brock); Knox, "migrant" (Rackliff); Oxford, (reported from this county in Smith's List of the Birds of Maine, Forest and Stream, Vol. 20, p. 66); Penobscot, "taken at Bangor, October 23, 1881" (Merrill, Bull. Nutt. Orn. Club, Vol. 7, p. 191); Washington, "rare" (Boardman).

107. (241). Tringa bairdii (*Coues*). Baird's Sandpiper.

A rare migrant along the coast and of accidental occurrence in the interior.

County Records.—Cumberland, "rare" (Brock); Knox, "rare visitant" (Norton); Oxford, "taken near Upton" (Brewster, Bull. Nutt. Orn. Club, Vol. 1, p. 191).

108. (242). Tringa minutilla *Vieill.* Least Sandpiper.

A very common migrant along the coast and of fairly common occurrence in the interior. A few birds are seen in midsummer along the coast. but it does not breed in the state. The southward migration begins in July and the birds are common through September. They arrive from the south in May.

County Records —Androscoggin, "common migrant" (Johnson); Cumberland, "common" (Brock); Hancock, "migrant" (Knight); Kennebec, (Dill); Knox, "summer" (Rackliff); Penobscot, (Hardy); Piscataquis, "migrant" (Homer); Sagadahoc, "common in summer" (Spinney); Somerset, "quite common migrant" (Morrell); Washington, "abundant in summer" (Boardman).

Subgenus PELIDNA Cuvier.

109. (243a). Tringa alpina pacifica (*Coues*). Red-backed Sandpiper.

A common autumn migrant along some parts of the coast, while on other parts it has not been met with by experienced observers. It is seemingly of rare occurrence in spring.

County Records.—Cumberland, "common" (Brock); Knox, "migrant" (Racklift); Washington, "rare" (Boardman).

Subgenus ANCYLOCHEILUS Kaup.

110. (244). Tringa ferruginea *Brünn.* Curlew Sandpiper.

Of accidental occurrence in the state, a specimen being killed at Pine Point, Cumberland County, on September 15, 1881, by Charles H. Chandler. (Cf. Purdie, Bull. Nutt. Orn. Club, Vol. 7, p. 124). It has also been taken over the boundary at Grand Menan, New Brunswick, by Mr. Boardman.

Genus EREUNETES Illiger.

111. (246). Ereunetes pusillus (*Linn.*). Semipalmated Sandpiper.

A very common migrant along the coast and somewhat rare in the interior. This species and the Least Sandpiper, with which it is often found associated, are the commonest species of the order, being found almost anywhere along the coast in the migrations. A few birds are sometimes seen in late July and by August they are common. It does not nest in the state.

County Records.—Cumberland, "rare near Bridgton" (Mead); "common" (Brock); Hancock, "abundant along the coast in migration" (Knight); Knox, "summer" (Racklift); Penobscot, (Hardy); Sagadahoc, "common in fall" (Spinney); Somerset, "quite common migrant" (Morrell); Washington, "common" (Boardman).

Genus CALIDRIS Cuvier.

112. (248). Calidris arenaria (*Linn.*). Sanderling.

Common along the coast in the fall migration, and of somewhat rarer occurrence in the spring.

County Records.—Cumberland, "common" (Brock); Knox, "migrant" (Racklift); Penobscot, "a specimen was killed at High Head, near Bangor, some years ago" (Crosby); Sagadahoc, "common in fall" (Spinney); Washington, "common" (Boardman).

Genus LIMOSA Brisson.

113. (249). Limosa fedoa (*Linn.*). Marbled Godwit.

Owing to the absence of any more definite evidence of the occurrence of this species as a regular visitor, it will have to be ranked as accidental. A specimen was taken at Scarborough Marsh, Cumberland County, in May, 1884, and is recorded by Brown in The Auk, Vol. 2, p. 385.

114. (251). Limosa hæmastica (*Linn.*). Hudsonian Godwit.

A rare autumn migrant along the coast where it is seemingly of somewhat local occurrence.

County Records.—Cumberland, "transient in autumn, generally rare" (Brown's Cat. Birds of Portland, p. 28); Washington, "rare" (Boardman).

Genus TOTANUS Bechstein.

115. (254). Totanus melanoleucus (*Gmel.*). Greater Yellow-legs.

A common migrant throughout the state, breeds in the north. It appears from the north early in August. Some years birds have been observed in July, but they are not known to have nested in the state.

County Records.—Androscoggin. "fairly common migrant" (Johnson); Cumberland, "rare near Bridgton" (Mead), "common" (Brock); Franklin, "rare migrant" (Richards); Hancock, "common migrant" (Dorr); Kennebec, "they have been shot here in abundance" (Powers); Knox, "summer" (Racklift); Oxford, "migrant" (Nash); Penobscot, "occurs in fall and spring" (Knight); Piscataquis, "migrant" (Homer); Sagadahoc, "common in fall" (Spinney); Somerset, "common migrant, birds were seen during the summer months of 1895" (Morrell); Washington, "common spring and fall" (Boardman).

116. (255). Totanus flavipes (*Gmel.*). Yellow-legs.

A common autumn and rare spring migrant throughout the state.

County Records. — Androscoggin, (Pike); Cumberland, "common" (Brock); Knox, (Racklift); Oxford, "not common at Norway")Verrill's List of the Birds of Norway); Penobscot, (Hardy); Sagadahoc, "common in fall" (Spinney); Somerset, "migrant" (Morrell); Washington, "common only in fall" (Boardman).

Subgenus HELODROMAS Kaup.

117. (256). Totanus solitarius (*Wils.*). Solitary Sandpiper.

A common migrant throughout the state. A few individuals remain through the summer in the northern counties, and while they doubtless breed I cannot positively state that such is the case.

County Records. — Androscoggin, "common migrant" (Johnson); Cumberland, "common" (Brock); Franklin, "common migrant" (Richards); Kennebec, "quite common" (Gardiner Branch); Knox, "summer" (Rackliff); Oxford, "not common at Norway" (Verrill's List of the Birds of Norway); Penobscot, "common migrant, several individuals were seen on Pushaw Stream, June 8, 1894, and also seen in the same localities in summers of 1895 and 1896" (Knight); Piscataquis, "rare" (Homer); Sagadahoc, "not plenty, a few in early fall" (Spinney); Somerset, "common migrant"(Morrell); Washington,"common" (Boardman).

Genus SYMPHEMIA Rafinesque.

118. (258). Symphemia semipalmata (*Gmel.*). Willet.

A rare migrant along the coast, chiefly occurring in late summer and autumn.

County Records.—Cumberland, "rare" (Brock); Knox, "rare in summer" (Rackliff); Sagadahoc, "quite rare, one taken October 25, 1896" (Spinney); Washington, "rare" (Boardman).

Genus PAVONCELLA Leach.

119. (260). Pavoncella pugnax (*Linn.*). Ruff.

An accidental visitor from Europe, of which two specimens have been taken in the state, and a third one just over the boundary at Grand Menan, by Mr. Boardman.

County Records.—Cumberland, "Scarborough, April 10, 1870" (Smith's List of the Birds of Maine, Forest and Stream, Vol. 20, p. 85); Oxford, "one taken September 8, 1874, at Upton" (Brewster, Bull. Nutt. Orn. Club, Vol. 1, p. 19).

Genus BARTRAMIA Lesson.

120. (261). Bartramia longicauda (*Bechst.*). Bartramian Sandpiper.

A common migrant throughout the state and quite a common summer resident in the interior counties.

4

County Records.—Androscoggin, "fairly common summer resident" (Johnson); Cumberland, "common summer resident" (Mead); Franklin, "common summer resident" (Swain); Kennebec, "rare" (Robbins); Knox, "rare visitant" (Norton); Oxford, "common, breeds" (Nash); Penobscot, "common migrant, rare summer resident" (Knight); Piscataquis, "common, breeds" (Homer); Sagadahoc, "a few in early fall" (Spinney); Somerset, "quite common summer resident" (Morrell); Washington, "accidental" (Boardman).

Genus TRYNGITES Cabanis.

121.　(262).　Tryngites subruficollis (*Vieill.*).　Buff-breasted Sandpiper.

Accidental along the coast. In the Forest and Stream, Vol. 20, p. 85, Mr. Smith records a specimen taken at Scarborough, Cumberland County. He has recently given information regarding a specimen which was taken at Cape Elizabeth in the above county, on September 13, 1887.

Genus ACTITIS Illiger.

122.　(263).　Actitis macularia (*Linn.*).　Spotted Sandpiper.

A very common summer resident throughout the state. It breeds both on the outer islands of the coast and along the ponds and streams of the interior, the eggs being laid in late May and early June. When flushed it utters a "peet-weet, peet-weet" and flies out over the water, and then in a semi-circular course back to the shore again. If followed up it will do this for a number of times, but finally instead of being driven further along the shore it will circle for a short distance back toward the place it was first driven from and again seek the shore. When on land it seems very uneasy and is constantly tipping, bowing, and tetering. From this habit it is locally known as Teter-up, Tip-tail, Tip-up, etc.

County Records.—Androscoggin, "common summer resident" (Johnson); Aroostook, "Fort Fairfield and Houlton, common" (Batchelder, Bull. Nutt. Orn. Club, Vol. 7, p. 151); Cumberland, "common summer resident" (Mead); Franklin, "common summer resident" (Swain); Hancock, "breeds commonly on the islands along the coast" (Knight); Kennebec, "common summer resident" (Gardiner Branch); Knox, "summer" (Rackliff); Lincoln, "common, breeding on the islands" (Norton); Oxford, "breeds commonly" (Nash); Penobscot, "very common summer resident" (Knight); Piscataquis, "common, breeds" (Homer); Sagadahoc, "common breeder" (Spinney); Somerset, "common summer resident" (Morrell); Waldo, "common" (Knight); Washington, "abundant summer resident" (Boardman); York, "common summer resident" (Adams).

Genus NUMENIUS Brisson.

123. (264). Numenius longirostris *Wils*. Long-billed Curlew.

A casual visitor in spring and late summer.

County Records.—Cumberland, "one at Scarborough, May 2nd, 1866" (Smith's List of the Birds of Maine, Forest and Stream, Vol. 20, p. 85); Knox, "rare visitant in summer" (Norton); Sagadahoc, "very rare in August" (Spinney); Washington, "very rare" (Boardman).

124. (265). Numenius hudsonicus *Lath*. Hudsonian Curlew.

A quite rare migrant along the coast in spring and autumn, breeding in the far north.

County Records.—Cumberland, "common" (Brock); Knox, "migrant" (Rackliff); Washington, "very rare" (Boardman).

125. (266). Numenius borealis (*Forst*.). Eskimo Curlew.

A migrant of varying abundance, chiefly occurring along the coast and less commonly in the interior of the state. In other parts of the country it is said to be far commoner in the interior than it is coastwise.

County Records.—Androscoggin, (Pike); Cumberland, "common" (Brock); Knox, "migrant" (Rackliff); Piscataquis, "rare visitor" (Homer); Sagadahoc, "very few in Aug." (Spinney); Washington, "very rare" (Boardman).

Family CHARADRIIDÆ. Plovers.

Genus SQUATAROLA Cuvier.

126. (270). Squatarola squatarola (*Linn*.). Black-bellied Plover.

Common along the coast in migrations and of rare occurrence in the interior.

County Records.—Androscoggin, (Pike); Cumberland, "common" (Brock); Hancock, "migrant" (Dorr); Knox, "migrant" (Rackliff); Lincoln, "as seen on Western Egg Rock, June 24, 1895" (Norton); Penobscot, "a specimen killed on Sunkhaze Stream is in the University of Maine collection" (Knight); Sagadahoc, "common fall and spring" (Spinney); Washington, "not very common" (Boardman).

Genus CHARADRIUS Linnæus.

127. (272). Charadrius dominicus *Mull*. American Golden Plover.

A common autumn migrant throughout the state but seemingly not occurring in the spring. This absence of the species in spring is due to their seeking their northern breeding grounds by a different route from that pursued in their journey southward.

County Records.—Androscoggin, "fairly common migrant" (Johnson); Cumberland, "common" (Brock); Kennebec, (Dill); Knox, (Rackliff); Oxford, (given in Verrill's List of the Birds of Norway); Penobscot, "common some falls and rare others" (Knight); Piscataquis, "migrant" (Homer); Sagadahoc, "rare in fall" (Spinney); Somerset, "two specimens shot by H. H. Johnson, September 10, 1894" (Morrell); Washington, "not very common" (Boardman); York, (Butters).

Genus ÆGIALITIS Boie.
Subgenus OXYECHUS Reichenbach.

128. (273). Ægialitis vocifera (*Linn.*). Killdeer.

A very rare migrant throughout the state, but still of such comparatively frequent occurrence as to prevent its being called accidental.

County Records.—Androscoggin, (Pike); Cumberland, "rare" (Brock); Knox, "rare migrant" (Rackliff); Penobscot, "used to occur here forty years ago" (Hardy); Piscataquis, "rare" (Homer); Sagadahoc, "very scarce in fall" (Spinney); Washington, "accidental" (Boardman).

Subgenus Ægialitis Boie.

129. (274). Ægialitis semipalmata *Bonap.* Semipalmated Plover.

A very common migrant along the coast and of fairly common occurrence in the interior. A few individuals remain all summer along the coast but they do not breed in the state. They are commonly called Ring Necks by hunters.

County Records.—Androscoggin, "fairly common migrant" (Johnson); Cumberland, "common" (Brock); Hancock, "common in migrations, I saw a flock of four individuals at Saddleback Ledge on June 22d, 1896" (Knight); Kennebec, (Dill); Knox, "summer" (Rackliff); Oxford, "occurs at Lake Umbagog" (Brewster, Bull. Nutt. Orn. Club, Vol. 5, p. 60); Penobscot, "taken at Stillwater by John Lord" (Knight); Sagadahoc, "common in August" (Spinney); Somerset, "one shot August 10th, 1894" (Morrell); Washington, "common in summer" (Boardman).

130. (277). Ægialitis meloda (*Ord*). Piping Plover.

Formerly a rare summer resident along our coast, probably still occurs in limited numbers. Its rarity is proved by the fact that several good observers have failed to detect this species of late years.

County Records.—Cumberland, "rare summer resident" (Brown's Cat. Birds of Portland, p. 25); Sagadahoc, "very scarce in August," (Spinney); Washington, "rare, said to breed on the islands" (Boardman).

131. (277a). Ægialitis meloda circumcincta *Ridgw.* Belted Piping Plover.

Of accidental occurrence along the coast in migrations. Only one specimen has been recorded from the state, this being taken at Scarborough, Cumberland County. (Cf. Allen, Auk, Vol. 3, p. 82).

Family APHRIZIDÆ. Surf Birds and Turnstones.

Subfamily ARENARIINÆ. Turnstones.

Genus ARENARIA Brisson.

132. (283). Arenaria interpres (*Linn.*). Turnstone.

Common along the coast in the migrations, occurring in May, late July, August, and September.

County Records.—Cumberland, "common" (Brock); Hancock, "found it common among the islands in August, 1896" (Knight); Knox, "summer" (Rackliff); Penobscot, "I am informed by Mr. Fred Colby, who is well informed regarding our birds, that he shot a specimen of this bird at Hermon Pond, its occurrence there being accidental" (Knight); Sagadahoc, "common in August" (Spinney); Washington, "fall, not rare" (Boardman).

Order GALLINÆ. Gallinaceous Birds.

Suborder PHASIANI. Pheasants, Grouse, Partridges, Quails, etc.

Family TETRAONIDÆ. Grouse, Partridges, etc.

Subfamily PERDICINÆ. Partridges.

Genus COLINUS Lesson.

133. (289). Colinus virginianus (*Linn.*). Bob-white.

Formerly a permanent resident of the southern part of the state, and while it still occurs, it is very rare. Under the date of July 12th, 1897, Mr. J. C. Mead of North Bridgton, Cumberland County, writes: "We have had a Bob-white with us now for nearly a month. Every now and then it comes into the pasture in the rear of my house and near the lake, and calls by the hour. Mrs. Mead and I had an excellent opportunity to watch it through the "glasses" at short range for a long time." This is the most recent record for the state.

County Records.—Androscoggin, "rare" (Johnson); Cumberland, "occasional" (Brock), "within past few years sportsmen have attempted to add it to our local game birds" (Brown's Cat. Birds of Portland, p. 38); Franklin, "very rare, not seen since 1890, and then only two seen" (Swain); Penobscot, "introduced to the county, some birds were let loose near Hermon, in 1894, by George Abbott, and these are reported to have bred in 1895 and 1896" (Knight).

Subfamily TETRAONINÆ. Grouse.
Genus DENDRAGAPUS Elliot.
Subgenus CANACHITES Stejneger.

134. (298). Dendragapus canadensis (*Linn.*). Canada Grouse.

This species is a very rare resident of the counties included in the Canadian fauna and probably breeds wherever found. In habits it is a tame and unsuspecting bird and can be easily shot or even knocked on the head with a club.

County Records.—Aroostook, "found at Houlton" (Batchelder in Bull. Nutt. Orn. Club, Vol. 7, p. 151); Franklin, "rare resident" (Richards); Hancock, "rare" (Dorr); Knox, "rare" (Racklift); Oxford, "breeds rarely" (Nash); Penobscot, "very rare resident" (Knight); Piscataquis, "resident, not common" (Homer); Somerset, "resident in northern part of county" (Morrell); Washington, "common resident" (Boardman).

Genus BONASA Stephens.

135. (300 a). Bonasa umbellus togata (*Linn.*). Canadian Ruffed Grouse.

While all the general and county lists of this state which have been published unite in calling our bird *bonasa umbellus* (Linn.), there seems to be much doubt that the typical Ruffed Grouse has ever been taken in the state. (Cf. Norton, Maine Sportsman, Vol. 4, No. 38, p. 6). Nevertheless Mr. Norton thinks that birds referable to *umbellus* will ultimately be taken in our southern counties. During the fall of 1896, hundreds of Grouse were examined in the Bangor markets by the editor, and specimens approaching *umbellus* were purchased and preserved for determination. Later these birds together with a series from Cumberland County, which were kindly furnished by Mr. J. C. Mead, were sent to Mr. William Brewster for identification, and he referred all to the race *togata*. Evidence bearing on the matter has been solicited from Ornithologists throughout the state, and nearly all have pronounced the birds of their locality to belong to this latter race. In view of this evidence, it has been deemed advisable to place the true Ruffed Grouse in our hypothetical list until its presence has been satisfactorily demonstrated by the production of specimens taken within our limits. The "Partridge" is a common resident throughout the state, breeding from early May to the middle of June. Perhaps the eggs may be deposited at an earlier date in some localities, but when fresh eggs are found at a later date than June 15th it is reasonably fair to assume that they are a second set, and doubtless due to the bird having been robbed of her first laying. The drumming of the cock bird is a sound well known to nearly every inhabitant of the state. While during the breeding season it probably serves to attract the female, yet it cannot be used exclusively for this purpose, as I have heard birds drumming in the late fall and even in midwinter. However, it is a fact that the birds drum most frequently in the spring and early summer. This drumming and the attitude assumed while doing so are admirably described by Mr. J. C. Mead in the Maine Sportsman for June, 1896, p. 6, and also by "Penobscot" in the September issue of the same on p. 6. The editor has examined birds from Aroostook, Cumberland, Franklin, Hancock, Penobscot, and Waldo Counties, and found all to be referable to *togata*.

County Records.—Androscoggin, "common resident" (Call); Aroostook, "reported as common" (Knight); Cumberland, "common resident" (Mead); Franklin, "common resident" (Lee and McLain); Hancock, "common resident both inland and on many of the wooded islands along the coast" (Knight); Kennebec, "common resident" (Powers); Knox, "resident" (Rackliff); Oxford, "breeds common" (Nash); Penobscot, "common resident" (Knight); Piscataquis, "common resident" (Homer); Sagadahoc, "nests" (Spinney); Somerset, "common" (Morrell); Waldo, "common" (Knight); Washington, "common" (Boardman); York, "would be common if sportsmen would let them alone" (Adams).

Genus LAGOPUS Brisson.

136. (301). Lagopus lagopus (*Linn.*). Willow Ptarmigan.

Of accidental occurrence at Kenduskeag, Penobscot County, where a specimen was shot on April 23, 1892. (Cf. Merrill, Auk Vol. 9, p. 300).

Family PHASIANIDÆ. Pheasants, etc.

Subfamily MELEAGRINÆ. Turkeys.

Genus MELEAGRIS Linnæus.

137. (310). Meleagris gallopavo *Linn.* Wild Turkey.

This species is recorded as having formerly been found in southern Maine. (Cf. Allen, Bull. Nutt. Orn. Club, Vol. 1, p. 55). Evidences of its having formerly occurred on Mount Desert Island, Hancock County, are also given by Mr. Townsend. (Cf. Townsend, ibid. p. 60). There are no records of the occurrence of the Wild Turkey in recent years so we may rate it as long ago extinct within our limits. In some of the southern and western states it still occurs in comparative abundance, but as soon as a given region becomes settled up the Turkeys disappear, owing to the destructive proclivities of man.

Order COLUMBÆ. Pigeons.

Family COLUMBIDÆ. Pigeons.

Genus ECTOPISTES Swainson.

138. (315). Ectopistes migratorius (*Linn.*). Passenger Pigeon.

Formerly an abundant migrant throughout the state, now nearly extinct. A few straggling individuals are seen semi-occasionally, but the great flights of Pigeons which formerly occurred are now things of the past. Mr. C. M. Hoxie, the well known Foxcroft taxidermist, writes: "Replying to your favor of recent date in regard to the Passenger Pigeon, I will say that one was shot about one-half mile from Dexter by a Mr. Frank Rogers, on August 16th, 1896. It was evidently a stray bird." This is the most recent record I have been able to obtain of this species which formerly was abundant and bred in favored localities.

County Records.—Androscoggin, "rare migrant" (Johnson); Cumberland, "none in ten years" (Mead); Franklin, "rare migrant" (Richards); Hancock, "I am informed by old hunters that this species formerly occurred abundantly in this county, and also that they nested in great numbers, none have been seen for ten years" (Knight); Kennebec, (given in Hamlin's List of the Birds of Waterville, Report of the Secretary of Maine Board of Agriculture for 1865, pp. 168–173); Knox, "rare in summer" (Racklift); Oxford, "rare migrant, specimens observed by different persons in the fall of 1891" (Johnson); Penobscot, "formerly abundant, no records of late" (Knight); Piscataquis, "rare, last seen in 1884" (Homer); Washington, "not uncommon formerly, all gone now" (Boardman); York, "last seen in September, 1885" (Adams).

Genus ZENAIDURA Bonaparte.

139. (316). Zenaidura macroura (*Linn.*). Mourning Dove.

A very rare summer resident of the southern counties of the state, but there are, to my knowledge, no records of the species having been observed nesting within our limits.

County Records.—Androscoggin, (Pike); Cumberland, "probably a rare summer resident" (Brown's Cat. Birds of Portland, p. 23); Knox, "rare in summer" (Racklift); Penobscot, "it has occurred on the Maine State College campus in late summer" (Prof. F. L. Harvey); Sagadahoc, "scattering, spring and fall" (Spinney); Washington, "accidental" (Boardman).

Order RAPTORES. Birds of Prey.

Suborder SACORHAMPHII. American Vultures.

Family CATHARTIDÆ. American Vultures.

Genus CATHARTES Illiger.

140. (325). Cathartes aura (*Linn.*). Turkey Vulture.

An accidental visitor to the state, of which four specimens have been taken.

County Records.—Cumberland, "one at Standish in summer of 1874" (Smith, Forest and Stream, Vol. 20, p. 26); Oxford, "one at East Fryeburg" (R. A. Gushee, Forest and Stream for 1883, p. 245); Penobscot, "one seen near Bangor, at Whitney's Hill, he sat a long time with his wings stretched up above his head, as the Eagle is represented on the "buzzard dollar"; I have seen hundreds of them so there is not a shade of doubt as to the identity" (Hardy); Washington, "very rare, one specimen" (Boardman); York, "one killed in Buxton, in December, 1876" (Brown's Cat. Birds of Portland, p. 23).

Genus CATHARISTA Vieillot.

141.   (326).   Catharista atrata (*Bartr.*).   Black Vulture.

This, like the preceding, occurs as an accidental visitor from the south.

County Records.—Oxford, "one at East Fryeburg" (Smith, Forest and Stream, Vol. 20, p. 285); Washington, "not uncommon some seasons" (Boardman); "Eastport" (Cf. Deane, Bull. Nutt. Orn. Club, Vol. 5, p. 63); "Calais" (Cf. Brewster, Auk, Vol. 10, p. 82).

Suborder FALCONES, Vultures, Falcons, Hawks, Buzzards, Eagles, Kites, Harriers, etc.

Family FALCONIDÆ.   Vultures, Falcons, Hawks, Eagles, etc.

Subfamily ACCIPITRINÆ.   Kites, Buzzards, Hawks, Goshawks, Eagles, etc.

Genus CIRCUS Lacépède.

142.   (331).   Circus hudsonius (*Linn.*).   Marsh Hawk.

A common summer resident of marshes and meadows.   The nest is placed on the ground in such localities.   The birds are common everywhere during the migrations, but during the breeding season they are only to be sought for near their favorite meadow or marsh. They will often return to the same locality for many successive seasons.

County Records.—Androscoggin, "common summer resident" (Johnson); Aroostook, "seen at Houlton and Fort Fairfield" (Batchelder, Bull. Nutt. Orn. Club, Vol. 7, p. 50); Cumberland, "common summer resident" (Mead); Franklin, "common summer resident" (Swain); Hancock, "summer resident" (Dorr); Kennebec, "rare" (Gardiner Branch); Knox, "summer" (Rackliff); Oxford, "breeds commonly" (Nash); Penobscot, "quite common summer resident" (Knight); Piscataquis, "common, breeds" (Homer); Sagadahoc, "common summer resident" (Spinney); Somerset, "common migrant, rare summer resident" (Morrell); Washington, "abundant, summer resident" (Boardman); Waldo, (Spratt); York, "breeds" (Adams).

Genus ACCIPITER Brisson.

Subgenus ACCIPITER.

143.   (332).   Accipiter velox (*Wils.*).   Sharp-shinned Hawk.

A common summer resident throughout the state, but owing to the difficulty of finding the nests, which are usually placed in evergreen trees, the eggs of this species continue to be objects of desiderata to collectors.

County Records.—Androscoggin, "common summer resident" (Johnson); Aroostook, "Houlton, not common" (Batchelder, Bull. Nutt. Orn. Club, Vol. 7, p. 151); Cumberland, "common summer resident" (Mead); Franklin, "common summer resident" (Swain); Hancock, "summer resident" (Dorr); Kennebec, "rare" (Gardiner Branch); Knox, "summer" (Rackliff); Oxford, "breeds commonly" (Nash); Penobscot, "breeds quite commonly" (Knight); Piscataquis, "common, breeds" (Homer); Sagadahoc, "common spring and fall" (Spinney); Somerset, "not common summer resident" (Morrell), "saw a flock of over a hundred migrating at Jackman, in August, 1895" (Harvey and Knight); Waldo, "breeds quite commonly" (Knight); Washington, "abundant summer resident" (Boardman); York, (Adams).

144. (333). Accipiter cooperii (*Bonap.*). Cooper's Hawk.

Of quite general occurrence as a summer resident throughout the state, but at the same time the species is by no means to be called common.

County Records.—Androscoggin, "common summer resident" (Johnson); Cumberland, "rare" (Mead); Franklin, "common summer resident" (Swain); Hancock, "summer resident" (Dorr); Kennebec, "rare" (Gardiner Branch); Knox, (Rackliff); Oxford, "summer resident" (Johnson); Penobscot, "seemingly quite a rare summer resident" (Knight); Piscataquis, "not uncommon" (Homer); Sagadahoc, "common migrant" (Spinney); Somerset, "rare summer resident" (Morrell); Washington, "not common, one of our rarest Hawks" (Boardman); York, (Adams).

Subgenus ASTUR Lacépède.

145. (334). Accipiter atricapillus (*Wils.*). American Goshawk.

Occurs commonly as a winter resident throughout the state, and less commonly as a resident of the portions within the Canadian fauna.

County Records.—Androscoggin, "common migrant" (Johnson); Cumberland, "common, have taken its eggs here" (Mead), "uncommon October to May" (Brown's Cat. Birds of Portland, p. 22); Franklin, "rare summer resident" (Swain); Hancock, "winter resident" (Dorr); Kennebec, (Gardiner Branch); Knox, "winter" (Rackliff); Oxford, "breeds rarely" (Nash); Penobscot, "common winter and quite rare summer resident, has been known to breed in several instances" (Knight); Piscataquis, "common resident" (Homer); Sagadahoc, "scattering, fall and spring" (Spinney); Somerset, "rare visitant" (Morrell); Washington, "not uncommon, breeds" (Boardman); York, (Adams).

Genus BUTEO Cuvier.

146. (337). Buteo borealis (*Gmel.*). Red-tailed Hawk.

Of quite general occurrence throughout the state in spring, summer, and autumn. A few are said to remain in the southern counties through the winter, but the majority retire southwards at the approach of cold weather.

County Records.—Androscoggin, "rare summer resident, common migrant" (Johnson); Aroostook, "common at Houlton" (Batchelder, Bull. Nutt. Orn. Club, Vol. 7, p. 151); Cumberland, "common resident" (Mead); Franklin, "common summer resident" (Swain); Hancock, "summer resident" (Dorr); Kennebec, "rare" (Gardiner Branch); Knox, "migrant" (Rackliff); Oxford, "breeds" (Nash); Penobscot, "a not uncommon summer resident" (Knight); Sagadahoc, "scattering, fall, winter, and spring" (Spinney); Waldo, "summer resident" (Spratt); Washington, "not uncommon summer resident, (Boardman); York, (Adams).

147. (339). Buteo lineatus (*Gmel.*). Red-shouldered Hawk.

This species is a fairly common summer resident, through most parts of the state. It has not been reported as wintering here.

County Records.—Androscoggin, "common summer resident" (Johnson); Cumberland, "common summer resident" (Mead); Franklin, "common summer resident" (Swain); Hancock, "summer resident" (Dorr); Kennebec, "common" (Powers); Knox, "migrant" (Rackliff); Oxford, "breeds rarely" (Nash); Penobscot, "a not uncommon summer resident" (Knight); Piscataquis, "rare" (Homer); Sagadahoc, "common migrant spring and fall" (Spinney); Somerset, "common summer resident" (Morrell;) Waldo, (Spratt); Washington, "not uncommon summer resident" (Boardman); York, "breeds" (Adams).

Subgenus TACHYTRIORCHIS Kaup.

148. (342). Buteo swainsoni *Bonap.* Swainson's Hawk.

An accidental visitor from the west, of which a number of specimens have been taken, all being melanistic.

County Records.—Hancock, "one taken at Gouldsborough September 15, 1886" (Cf. Brewster, Auk, Vol. 5, p. 424); Penobscot, "one at Glenburn, May 19, 1888" (Cf. Brewster, Ibid, Vol. 5, p. 424), "have seen at least two taken here, both melanistic" (Hardy); Washington, "taken at Calais, about October 8, 1892" (Cf. Brewster, Auk, Vol. 10, p. 82).

149. (343). Buteo latissimus (*Wils.*). Broad-winged Hawk.

A quite common summer resident in most portions of the state.

It arrives from the south early in April, and the eggs are usually deposited by the second or third week of May.

County Records.—Androscoggin, "fairly common summer resident" (Johnson); Aroostook, "breeding at Houlton" (Cf. Batchelder, Bull. Nutt. Orn. Club, Vol. 7, p. 151); Cumberland, "common summer resident" (Mead); Franklin, "rare summer resident" (Swain); Hancock, "summer resident" (Dorr); Kennebec, Royal); Knox, "migrant" (Rackliff); Oxford, "breeds commonly" (Nash); Penobscot, "breeds quite commonly, and is apparently the commonest of our larger Hawks" (Knight); Piscataquis, "common, breeds" (Homer); Sagadahoc, "common spring and fall" (Spinney); Somerset, "not common summer resident" (Morrell); Waldo, (Spratt); Washington, "abundant summer resident" (Boardman); York. (Adams .

Genus ARCHIBUTEO Brehm.

150. (347 a). Archibuteo lagopus sancti-johannis (Gmel.). American Rough-legged Hawk.

This species is of quite common occurrence as a winter visitor from the north. It does not breed in the state, all previous records stating to the contrary notwithstanding. In Bendire's Life Histories of North American Birds he states that it is not known to breed in the United States, save in Alaska.

County Records.—Androscoggin, "fairly common winter resident" (Johnson); Cumberland, "rare winter visitant" (Brown's Cat. Birds of Portland, p. 22 ; Franklin, "rare" (Richards); Knox, "winter" (Rackliff) Oxford, "common at Norway in winter" (Verrill's List of the Birds of Norway, Proc. Essex Institute, Vol. 3, pp. 136 et seq. ; Penobscot, "rare winter visitor" (Knight); Piscataquis, "rare" (Homer); Sagadahoc, "a few in fall and early winter" (Spinney); Washington, "very rare winter resident" (Boardman).

Genus AQUILA Brisson.

151. (349). Aquila chrysaëtos (*Linn.*). Golden Eagle.

Occasionally observed and taken here, but at present the evidence points to its occurrence only as a straggler. While it has been observed here in summer under circumstances that would hint that it might possibly nest in the wild, mountainous parts of the state, there has unfortunately been no positive proof brought forth to substantiate this belief. While at Jackman, in August, 1895, Prof. F. L. Harvey and myself saw what we are positive was one of these birds on Sandy Bay Mountain. This bird was seemingly uneasy at our presence and flew very near us uttering its shrill cry. The cry was answered from a steep cliff on the side of the mountain, so the bird either had a mate or young in the immediate vicinity. While the bird repeatedly approached near enough to us to render us *certain* in our own minds of its identity, we unfortunately had no gun with us, and so could not secure the bird to render its identity absolutely certain, as is demanded by modern science. Again, not seeing the bird which answered its cries we cannot state whether it was a mate to the one we saw or its young. In this connection see Knight, The Auk, Vol. 13, p. 82.

County Records.—Androscoggin, (Pike); Cumberland, "taken at Peak's Island" (Brown's Cat. Birds of Portland, p. 22), "specimens are also recorded by Dr. Brock" (Cf. Brock, Auk, Vol. 13, p. 256); Franklin, "rare" (Richards); Somerset, "while positive that we saw one at Sandy Bay Mountain, in August, 1895, the specimen was not killed so as to establish a record beyond doubt" (Harvey and Knight); Washington, "very rare, shot in summer" (Boardman).

Genus HALIÆETUS Savigny.

152. (352). Haliæetus leucocephalus (*Linn.*). Bald Eagle.

Resident throughout the year along the coast, where it is fairly common. It is also quite common as a summer resident throughout the interior, where it frequents the vicinity of ponds and lakes.

County Records.—Androscoggin, "fairly common summer resident" (Johnson); Aroostook, "not common at Houlton" (Batchelder, Bull. Nutt. Orn. Club, Vol. 7, p. 151); Cumberland, "summer resident" (Mead); Franklin, "rare summer resident" (Swain); Hancock, "resident along the coast, often seen about Isle au Haut, breeds" (Knight); "Kennebec, very rare" (Powers); Knox, "resident" (Rackliff); Oxford, "breeds rarely" (Nash); Penobscot, "a pair nest near Pushaw Pond every year, and they are of quite common occurrence in the vicinity of ponds and lakes throughout the county" (Knight); Piscataquis, "not uncommon, breeds" (Homer); Sagadahoc, "common and breeds" (Spinney); Somerset, "occasional visitant" (Morrell); Waldo, "rare" (Spratt; Washington, "common, breeds" (Boardman).

Subfamily FALCONINÆ. Falcons.

Genus FALCO Linnæus.

Subgenus HIEROFALCO Cuvier.

153. (353). Falco islandus *Brünn.* White Gyrfalcon.

An accidental winter visitor from the north, there is but one state record of its occurrence. This specimen is recorded by Mr. Brew-

ster, who says: "Not long since Mr. George A. Boardman wrote
me that he had heard of the capture, in eastern Maine, of a very
light-colored Gyrfalcon. Upon my expressing a lively interest in
the matter he very kindly put me in correspondence with Mr. John
Clayton of Lincoln, Maine, who mounted the specimen, and from
whom I have just purchased it. Although too dark-colored to be
typical of that form it is, nevertheless, an unmistakable example
of *F. islandus*, Brünn. It was shot in South Winn, about October
8, 1893, by a young man named Wyman, who found it perched on
a telegraph pole." (Cf. Brewster, The Auk, Vol. 12, p. 180).
This Penobscot County specimen is unique in Maine, and even the
United States, as is in fact stated by Mr. Brewster in the article
above quoted.

154. (354). Falco rusticolus *Linn*. Gray Gyrfalcon.

A specimen taken at Cape Elizabeth, October, 13th, 1877, and
recorded by Mr. Brown under the name of *Hierofalco gyrfalco
islandicus*, is the only New England example of this bird known to
exist. It is at present in the collection of Prof. J. Y. Stanton of
Lewiston. The record will be found in Brown's Catalogue of the
Birds of Portland, p. 21. In Minot's Land and Game Birds of
New England, 2nd edition, page 479, Mr. Brewster in speaking
of this specimen says: "I have not yet seen it, but if it has been
correctly identified it is the only New England example of
rusticolus of which I have any present knowledge." Mr. Boardman
writes me that he has at least two specimens of this bird, but
unfortunately they were both taken on Canadian soil, and though
taken near our boundary they are not Maine specimens.

155. (354a). Falco rusticolus gyrfalco (*Linn*.). Gyrfalcon.

But one instance of its occurrence in the state is known to me.
A specimen is recorded under the name of *Falco gyrfalco sacer*
(Forst.) as being taken at Ktaadn Iron Works, Piscataquis County,
December, 1876. (Cf. Purdie, Bull. Nutt. Orn. Club, Vol. 4, p.
188). After correspondence with Mr. Purdie who is positive of
the correctness of his identification, I have added the species to
our list as a casual winter visitor.

156. (354b). Falco rusticolus obsoletus (*Gmel*.). Black
Gyrfalcon.

Like others of its near relatives it ranks as a rare winter visitor,
though more abundant than the other Gyrfalcons.

County Records.—Cumberland, "Mr. E. P. Carman of Bridgton has a specimen which was shot between Cape Elizabeth and Lewiston about the middle of September, 1887" (information regarding this received from Prof. Stanton and Mr. Mead, the latter having kindly obtained the complete particulars regarding it); Knox, "one is recorded by Mr. Brewster as being taken near Rockland in November, 1886" (Cf. Brewster, The Auk, Vol. 4. p. 75), and as Mr. Rackliff of Spruce Head took a Gyrfalcon on the day before Thanksgiving at about that year, and sent it to F. B. Webster of whom Mr. B. secured his bird, it would seem assured that these birds are one and the same"; "in a recent letter from Mr. Brewster he informs me that he has another of these birds secured at Eagle Island, about March 22, 1888" (Cf. Brewster, Minot's Land and Game Birds of New England, 2nd Ed., p. 480, for the record of this specimen); Oxford, "one shot in 1892 or 1893" (Nash ; Washington, "winter visitant, three specimens" (Boardman).

Subgenus RHYNCHODON Nitzsch.

157. (356). Falco peregrinus anatum (*Bonap.*). Duck Hawk.

A quite rare summer resident of the isolated mountainous portions of the state and there found breeding. In migrations it is somewhat commoner, though still quite rare. It is quite likely to be found in winter in the southern parts of the state although not yet so reported.

County Records.—Androscoggin, (Pike); Cumberland, "very rare transient" (Brown's Cat. Birds of Portland, p. 21); Oxford, "rare, breeds" (Nash); Penobscot, "quite rare, may possibly breed" (Knight); Washington, not uncommon, breeds" (Boardman)."

Subgenus ÆSALON Kaup.

158. (357). Falco columbarius *Linn.* Pigeon Hawk.

Quite common everywhere in migrations, and a rare summer resident within the Canadian fauna. While the species occurs in summer I have not been able to ascertain that any nests or eggs have been found within our boundaries.

County Records.—Androscoggin, "summer" (Johnson); Cumberland, "not common" (Mead); Franklin, "rare summer resident, a specimen was shot here in June" (Swain); Kennebec, "rare" (Gardiner Branch); Knox, "migrant" (Rackliff); Oxford, "rare summer resident" (Nash); Penobscot, "common in spring and fall" (Knight); Piscataquis, "not uncommon" (Homer); Washington, "not uncommon summer resident" (Boardman); York, (Adams).

Subgenus TINNUNCULUS Vieillot.

159. (360). Falco sparverius *Linn.* American Sparrow Hawk.

Of quite general distribution as a summer resident throughout the interior of the state, but seemingly commonest during the migrations. It is not characteristic of any one faunal region, being found breeding from Florida to Hudson Bay.

County Records.—Androscoggin, "fairly common summer resident" (Johuson); Aroostook, "seen at Fort Fairfield" (Batchelder, Bull. Nutt. Orn. Club, Vol. 7, p. 15); Cumberland, "rare" (Mead), "have informa- ion of its breeding in Windham" (Norton); Franklin, "summer resi- dent" (Richards); Hancock, "summer resident" (Murch); Kennebec, "rare" (Gardiner Branch); Knox, "migrant" (Racklift); Oxford, "com- mou, breeds" (Nash); Penobscot, "quite common in migration, the eggs have been taken near Bangor" (Knight); Piscataquis, "common, breeds" (Homer); Sagadahoc, "common spring and fall" (Spinney); Somerset, "not common summer resident" (Morrell); Washington, "not uncommon summer resident" (Boardman); York, (Adams).

Subfamily PANDIONINÆ. Ospreys.

Genus PANDION Savigny.

160. (364). Pandion haliaëtus caroliuensis (*Gmel.*). Ameri- can Osprey.

The Fish Hawk is a common summer resident along the coast, and also quite common about the ponds and lakes of the interior. On the coast the nests are usually placed low down in the stunted trees that grow on the islands, and in many cases the ground itself serves as foundation. I have seen at least six nests placed thus, usually being situated on some rocky point of an island. The nests in trees, while usually near the shore of the islands, are quite often placed some distance inland. In such localities the birds are some- what social, and two or three nests are often placed within a radius of one hundred yards. About the interior ponds and lakes the birds are not so social, it being unusual to notice more than one pair of birds about a given body of water. They are also more wary in their selection of a nesting site, invariably choosing the tallest tree in the vicinity.

County Records.—Androscoggin, "common" (Johnson ; Cumberland, "common summer resident" (Mead); Franklin, "rare summer resident" (Swain); Hancock, "very common summer resident" (Knight); Kennebec, "rare" (Royal); Knox, "summer" (Rackliff); Oxford, "breeds rarely" (Nash); Penobscot, "fairly common summer resident, middle of April to September" (Knight); Piscataquis, "not uncommon, breeds" (Homer); Sagadahoc, "common summer resident" (Spinney); Somerset, "common, probably summer resident" (Morrell); Waldo, "common summer resident" (Knight); Washington, "abundant summer resident" (Boardman); York, (Adams).

Suborder STRIGES Owls.
Family BUBONIDÆ. Horned Owls, etc.
Genus ASIO Brisson.

161. (366). Asio wilsonianus (*Less.*). American Long-eared Owl.

A resident throughout the state, but still seemingly more numerous in fall than at other times. Perhaps this reported greater abundance in fall is due to the fact that people seek the haunts of this bird to hunt at this season, and consequently the birds are more likely to come under observation.

County Records.—Androscoggin, "fairly common resident" (Johnson); Cumberland, "not rare" (Mead); Franklin, "common resident" (Richards ; Hancock, "resident" (Dorr); Kennebec, "given in Hamlin's List of the Birds of Waterville" (See Report of Sec'y Me. Board of Agriculture, 1865, pp. 168-173); Knox, "migrant" (Rackliff); Oxford, "breeds rarely" (Nash); Penobscot, "somewhat rare, oftenest seen in fall" (Knight); Piscataquis, "not uncommon, breeds" (Homer; Sagadahoc, "common spring and fall" (Spinney); Washington, "not uncommon resident" (Boardman).

162. (367). Asio accipitrinus (*Pall.*). Short-eared Owl.

Resident in limited numbers, but oftenest occurring in fall or spring when it is common, especially along the coast, where it frequents the grassy salt marshes.

County Records.—Androscoggin, "fairly common resident" (Johnson); Cumberland, "moderately common resident" (Brown's Cat. Birds of Portland, p. 20); Franklin, "rare resident" (Richards); Hancock, "resident" (Dorr); Knox, "migrant" (Rackliff); Oxford, "rare" (Nash); Penobscot, "of rare occurrence in the fall" (Knight); Sagadahoc, "common spring and fall" (Spinney); Washington, "not uncommon resident" (Boardman); York, (Butters).

Genus SYRNIUM Savigny.

163. (368). Syrnium nebulosum (*Forst.*). Barred Owl.

This is the commonest of our larger Owls, being resident in wooded districts throughout the state. Like our other species it is most frequently observed in fall and winter.

County Records.—Androscoggin, "fairly common resident" (Johnson);
Aroostook, "occurs at Fort Fairfield and Houlton" (Batchelder, Bull.
Nutt. Orn. Club, Vol. 7, p. 50); Cumberland, "resident, rare in summer" (Brown's Cat. Birds of Portland, p. 20); Franklin, "common
resident" (Richards); Hancock, "common resident" (Dorr); Kennebec,
"common" (Powers); Knox, "winter" (Rackliff); Oxford, "breeds commonly" (Nash); Penobscot, "resident, commonest in fall" (Knight);
Piscataquis, "abundant. breeds" (Homer ; Sagadahoc, "common spring
and fall" (Spinney); Somerset, "not common resident" (Morrell);
Washington, "abundant resident" (Boardman).

Genus SCOTIAPTEX Swainson.

164. (370). Scotiaptex cinerea (*Gmel.*). Great Gray Owl.

An irregular winter visitor from the north. It does not occur
some seasons, and again it may be found fairly common at others,
but still, even when at its greatest abundance, it is a comparatively
rare bird.

County Records.—Androscoggin, (Pike); Cumberland, "six specimens
known additional to those recorded in Smith's List" (record from E.
Smith); Franklin,"very rare" (Swain); Hancock, "rare" (Dorr); Kennebec, "one at Augusta in December, 1887" (E. Smith); Knox, "winter"
(Rackliff); Oxford, "rare visitant" (Nash); Penobscot, "Mr. S. L. Crosby
informs me that these birds were quite common here one winter in the
early '90's" (Knight); Piscataquis, "rare winter visitor" (Homer);
Washington, "rare, winter only" (Boardman); York, "one taken at Biddeford, March 2d, 1890" (E. Smith).

Genus NYCTALA Brehm.

165. (371). Nyctala tengmalmi richardsoni (*Bonap.*). Richardson's Owl.

A somewhat irregular winter visitor, but still often not uncommon
locally at this season.

County Records.—Androscoggin, "fairly common winter visitant"
(Johnson); Cumberland, "rare" (Mead); Franklin, "rare winter resi-
dent" (Swain); Hancock, "rare" (Dorr): Kennebec, "very rare" (Dill);
Knox, "rare in winter" (Rackliff); Lincoln, "taken at Waldoborough"
(Smith, Forest and Stream, Vol. 20, p. 285); Oxford, "rare" (Nash);
Penobscot, "a quite rare and irregular winter visitor" (Knight); Piscat-
aquis, "rare winter visitor" (Homer); Sagadahoc, "taken at Bath" (C.
H Greenleaf to E. Smith); Washington, "not uncommon in winter"
(Boardman).

166. (372). Nyctala acadica (*Gmel.*). Saw-whet Owl.

Resident throughout the state, but on account of its frequenting somewhat low swampy woods in summer it is less often noticed at this season. In fall and winter it is quite common. The nests and eggs seem to have been taken more often in Franklin County than in any other part of the state. It is known to many as the Acadian Owl.

County Records.—Androscoggin, "rare resident" (Johnson); Aroostook, "common at Fort Fairfield and Houlton" (Batchelder, Bull. Nutt. Orn. Club, Vol. 7, p. 150); Cumberland, "common" (Mead); Franklin, "common resident" (Richards); Kennebec, "rare" (Gardiner Branch); Knox, "resident" (Rackliff); Oxford, "breeds" (Nash); Penobscot, "I have observed it only in fall and winter, but am of the opinion that it must occur in summer also" (Knight); Piscataquis, "common resident, breeds" (Homer); Sagadahoc, "spring, fall and winter" (Spinney); Somerset, "resident" (Morrell); Waldo, (Spratt); Washington, "common resident" (Boardman); York, "breeds" (Adams).

Genus MEGASCOPS Kaup.

167. (373). Megascops asio (*Linn.*). Screech Owl.

A rare resident of those counties in the Canadian fauna, somewhat commoner in those of the Alleghanian.

County Records.—Androscoggin, "common resident" (Johnson); Cumberland, "rare" (Mead); Franklin, "rare resident" (Richards); Hancock, "rare resident" (Dorr); Kennebec, "very rare" (Gardiner Branch); Knox, "migrant" (Rackliff); Oxford, "breeds" (Nash); Penobscot, "rare" (Knight); Piscataquis, "rare" (Homer); Sagadahoc, "one specimen in late fall" (Spinney); Somerset, "resident" (Morrell); Waldo, (Spratt); Washington, "very rare" (Boardman); York, "breeds" (Adams).   •

Genus BUBO Duméril.

168. (375). Bubo virginianus (*Gmel.*). Great Horned Owl.

A common resident of wooded districts throughout the state, but like others of this Suborder it is most often noticed in fall and winter.

County Records.—Androscoggin, "fairly common resident" (Johnson); Aroostook, "common at Houlton and Fort Fairfield" (Batchelder, Bull. Nutt. Orn. Club, Vol. 7, p. 150); Cumberland, "common resident" (Mead); Franklin, "common resident" (Swain); Hancock, "common resident" (Dorr); Kennebec, "rare" (Gardiner Branch); Knox, "resident" (Rackliff); Oxford, "breeds commonly" (Nash); Penobscot, "fairly common resident" (Knight); Piscataquis, "common resident" (Homer); Sagadahoc, "common, nests" (Spinney); Somerset, "not common resident" (Morrell); Waldo, (Knight); Washington, "common resident" (Boardman); York, "breeds" (Adams).

169. (375b). Bubo virginianus arcticus (*Swains.*). Arctic Horned Owl.

A specimen of this subspecies is at present in the collection of the Portland Society of Natural History where I have recently had the pleasure of examining it. This is recorded in the proceedings of the above society for April 1st, 1897 by Mr. A. H. Norton who writes as follows: "The collection of the Portland Society of Natural History contains an Owl strongly characteristic of this subspecies which was given to the society alive by Mr. Sewall Cloudman, December 6th, 1869. The only locality recorded is Maine." A second specimen is reported to me by Mr. J. Waldo Nash of Norway who writes: "I mounted an Arctic Horned Owl in 1886 that was shot in Brownfield." Its occurrence is probably casual or accidental. I have recently seen in the collection of Mr. Geo. A. Boardman of Calais a typical example of this subspecies which was taken just over the line in New Brunswick.

170. (375c). Bubo virginianus saturatus *Ridgw.* Dusky Horned Owl.

Admitted to the list upon the strength of a specimen in the collection of the Portland Society of Natural History, recorded by Mr. A. H. Norton who writes: "The collection contains a Dusky Horned Owl in which the characteristics of the race are strongly marked. It was given by Dr. Benjamin F. Fogg, March 12th, 1870, when it was recorded as a fresh specimen. Though the locality is recorded as Maine there are reasons for the belief that this and the specimen of *arcticus* were taken near Portland." (Cf. Norton, Proc. Port. Soc. Nat. Hist., Apr. 1st, 1897, p. 103). I have recently had the privilege of viewing this specimen also at the Society's rooms in Portland. The evidence now at hand leads me to believe that this species occurs casually in the state.

Genus NYCTEA Stephens.

171. (376). Nyctea nyctea (*Linn.*). Snowy Owl.

A regular winter visitor to the state and, while usually rare, it sporadically occurs in comparative abundance. Although of very general occurrence throughout the state, it is apt to be found in greater numbers along the coast.

County Records.—Androscoggin, "rare winter visitor" (Johnson); Cumberland, "rare" (Mead); Franklin, "rare winter visitor" (Swain) Hancock, "rare" (Dorr); Kennebec, "very rare" (Powers); Knox, "winter" (Rackliff); Oxford, "very rare" (Nash); Penobscot, "rare, one seen in January, 1895" (Knight); Piscataquis, "rare winter visitor" (Homer); Sagadahoc, "irregularly common winter visitor" (Spinney); Washington, "uncertain, some winters common" (Boardman); York, (Adams).

Genus SURNIA Duméril.

172. (377a). Surnia ulula caparoch (_Müll._). American Hawk Owl.

Occurs as an irregularly common winter visitor of quite general distribution.

County Records.—Androscoggin, "rare winter visitor" (Johnson); Aroostook, "occurs at Houlton" (Cf. Brewer, Bull. Nutt. Orn. Club, Vol. 2, p. 78); Cumberland, "rare" (Mead); Franklin, "rare" (Richards); Hancock, "rare" (Dorr); Kennebec, "very rare" (Royal); Knox, "migrant" (Rackliff); Oxford, "rare" (Nash); Penobscot, "quite common some seasons in late fall and winter" (Knight); Piscataquis, "some winters common" (Homer); Washington, "some winters common" (Boardman).

Order COCCYGES. Cuckoos, etc.

Suborder CUCULI. Cuckoos, etc.

Family CUCULIDÆ. Cuckoos, Anis, etc.

Subfamily COCCYGINÆ. American Cuckoos.

Genus COCCYZUS Vieillot.

173. (387). Coccyzus americanus (_Linn._). Yellow-billed Cuckoo.

A very rare summer resident of those counties within the Alleghanian fauna, while elsewhere it must be ranked as accidental or casual.

County Records.—Androscoggin, "rare summer resident" (Stanton); Cumberland, "rare summer resident" (Brown's Cat. Birds of Portland, p. 20); Hancock, "one taken at Bar Harbor by E. Gordon" (Smith); Oxford, "visitant" (Nash); Washington, "accidental" (Boardman).

174. (388). Coccyzus erythropthalmus (_Wils._). Black-billed Cuckoo.

A common summer resident, but on account of its somewhat retiring habits it is a bird more often heard than seen. Feeding to a large extent upon the larvæ of the "Tent Moth" and other injurious catapillars, it is one of the most beneficial of our birds.

County Records.—Androscoggin, "common summer resident" (Johnson); Aroostook, "occurs at Houlton" (Batchelder, Bull. Nutt. Orn. Club, Vol. 7, p. 150;) Cumberland, "common summer resident" (Mead); Franklin, "common summer resident" (Swain); Hancock, "summer resident" (Dorr); Kennebec, "common summer resident" (Royal); Knox, "summer" (Rackliff); Oxford, "breeds common" (Nash); Penobscot, "common summer resident" (Knight); Piscataquis, "common, breeds" (Homer); Sagadahoc, "common summer resident" (Spinney); Somerset, "common summer resident" (Morrell); Waldo, (Spratt); Washington, "common" (Boardman); York, "abundant breeder" (Adams).

Suborder ALCYONES. Kingfishers.
Family ALCEDINIDÆ. Kingfishers.
Genus CERYLE Boie.
Subgenus STREPTOCERYLE Bonaparte.

175.   (390).   Ceryle alcyon (Linn.).   Belted Kingfisher.

A common summer resident, and while it is usually found in the vicinity of water, I have found nests situated at least a mile from the nearest stream or brook. The nest is always situated in a burrow dug in the perpendicular face of a sand bank, and the excavation varies in length from three to twelve feet.

County Records.—Androscoggin, "common summer resident" (Johnson); Aroostook, "rather common at Fort Fairfield" (Batchelder, Bull. Nutt. Orn. Club, Vol. 7, p. 150); Cumberland, "common summer resident" (Mead); Franklin, "common summer resident" (Lee and McLain); Hancock, "summer resident" (Murch), "breeds on Deer Isle" (Knight); Kennebec, "common summer resident" (Gardiner Branch); Knox, "summer" (Rackliff); Oxford, "common, breeds" (Nash); Penobscot, "breeds commonly along the Penobscot River and elsewhere throughout the county" (Knight); Piscataquis, "common, breeds" (Homer); Sagadahoc, "common summer resident" (Spinney); Somerset, "common summer resident" (Morrell); Waldo, "summer resident" (Knight); Washington, "abundant summer resident" (Boardman); York, "common on Saco River" (Adams).

Order PICI. Woodpeckers, Wrynecks, etc.

Family PICIDÆ. Woodpeckers.

Genus DRYOBATES Boie.

176. (393). Dryobates villosus (*Linn.*). Hairy Woodpecker.

One of the commonest of its family throughout the state, being exceeded in numbers only by the Flicker and Downy Woodpecker. It is resident wherever found within our limits, in summer retiring to the solitudes of the country to nest, while in winter it is common in the trees and orchards of our city gardens.

County Records.—Androscoggin, "fairly common resident" (Johnson); Aroostook, "common at Fort Fairfield" (Batchelder, Bull. Nutt. Orn. Club, Vol. 7, p. 150); Cumberland, "common resident" (Mead); Franklin, "common resident" (Lee and McLain); Hancock, "resident" (Murch); Kennebec, "common resident" (Gardiner Branch); Knox, "resident" (Rackliff); Oxford, "common, breeds" (Nash); Penobscot, "breeds quite commonly, especially common in winter when it may be observed almost daily in the heart of the city of Bangor" (Knight); Piscataquis, "common resident" (Homer); Sagadahoc, "common, nests" (Spinney); Somerset, "common resident" (Morrell); Waldo, (Spratt); Washington, "abundant" (Boardman); York, "quite common" (Adams).

177. (394c). Dryobates pubescens medianus (*Swains.*). Downy Woodpecker.

According to Mr. Brewster (Cf. Brewster, Auk, Vol. 14, p. 82) our northern Downy Woodpecker is subspecifically separable from the southern bird, and as the type of D. pubescens came from the south, our northern bird will become a subspecies for which Dryobates pubescens medianus (Swains.), will become the first available name, according to the rule of priority in nomenclature. As the assigned habitat of this race is "Middle and northern parts of eastern United States and northward" all records of our Maine Downy Woodpecker will refer to this subspecies. It is a common resident throughout the state.

County Records. — Androscoggin, "common resident" (Johnson); Aroostook, "common at Fort Fairfield" (Batchelder, Bull. Nutt. Orn. Club, Vol. 7, p. 150); Cumberland, "common resident" (Mead); Franklin, "common resident" (Swain); Hancock, "common resident, nests very commonly on the wooded islands along the coast" (Knight); Kennebec, "abundant resident"(Gardiner Branch); Knox, "resident" (Rackliff); Oxford, "common, breeds" (Nash); Penobscot, "next to the Flicker it is our commonest Woodpecker here, and it is the commonest resident species" (Knight); Piscataquis, "common resident" (Homer); Sagadahoc, "common fall and spring" (Spinney); Somerset, "common resident" (Morrell); Waldo, (Spratt); Washington, "abundant" Boardman); York, "quite common" (Adams).

Genus PICOIDES Lacépède.

178. (400). Picoides arcticus (*Swains.*). Arctic Three-toed Woodpecker.

A rare summer resident of the extreme northern and eastern counties, while it is a fairly common winter visitant throughout nearly the entire state.

County Records.—Androscoggin, "rare winter visitor" (Johnson); Aroostook, "seen at Fort Fairfield" (Batchelder, Bull. Nutt. Orn. Club, Vol. 7, p. 150); Cumberland, "rare winter visitant" (Mead); Franklin, "rare" (Richards); Hancock, "in winter" (Dorr); Kennebec, "very rare" (Powers); Knox, "migrant" (Rackliff); Oxford, "breeds rarely' (Nash); Penobscot, "not uncommon in late fall and winter" (Knight); Piscataquis, "common in winter" (Homer); Somerset, "one specimen taken February 9, 1895" (Morrell); Waldo, "rare" (Spratt); Washington, "not uncommon in winter, rare summer resident" (Boardman).

179. (401). Picoides americanus *Brehm:* American Three-toed Woodpecker.

An exceedingly rare resident of the Canadian fauna, and a rare winter visitant elsewhere in the state. Reported as resident in two counties only.

County Records—Franklin, "rare resident" (Richards); Kennebec, "very rare" (Powers); Oxford, "winter visitant" (Nash); Penobscot, "very rare" (Hardy); Piscataquis, "rare winter visitor" (Homer); Sagadahoc, "rare, two specimens only" (Spinney); Washington, "not uncommon, rare summer resident" (Boardman).

Genus SPHYRAPICUS Baird.

180. (402). Sphyrapicus varius (*Linn.*). Yellow-bellied Sapsucker.

A common summer resident of most parts of the state. It is found from late March and early April to late in September.

County Records.—Androscoggin, "fairly common summer resident" (Johnson); Aroostook, "commonest Woodpecker at Fort Fairfield" (Batchelder, Bull. Nutt. Orn. Club, Vol. 7, p. 150); Cumberland, "common summer resident" (Mead); Franklin, "common summer resident" (Swain); Hancock, "common summer resident" (Knight); Kennebec, "common summer resident" (Sanborn); Knox, "migrant" (Rackliff); Oxford, "breeds commonly" (Nash); Penobscot, "quite common summer resident, very common migrant" (Knight); Piscataquis, "abundant summer resident" (Whitman); Sagadahoc, "common migrant" (Spinney); Somerset, "common summer resident" (Morrell); Waldo, "rare" (Spratt); Washington, "common summer resident" (Boardman).

Genus CEOPHLŒUS Cabanis.

181. (405). Ceophlœus pileatus (*Linn.*). Pileated Woodpecker.

Formerly quite common, but now its center of abundance is coincident with the heavily timbered and unsettled portions. It is resident, and breeds wherever found throughout the wilder parts of the state.

County Records.—Androscoggin, "rare resident" (Johnson); Aroostook, "common at Houlton" (Batchelder, Bull. Nutt. Orn. Club, Vol. 7, p. 150); Cumberland, "common resident" (Mead); Franklin, "rare resident" (Swain); Hancock, "resident" (Murch); Kennebec, "very rare resident" (Dill); Oxford, "breeds rarely (Nash); Penobscot, "fairly common resident in the unsettled parts of the county" (Knight); Piscataquis, "common resident" (Homer); Somerset, "rare resident" (Morrell); Waldo, "rare" (Spratt); Washington, "not uncommon resident" (Boardman); York, "a few seen yearly" (Adams).

Genus MELANERPES Swainson.

Subgenus MELANERPES.

182. (406). Melanerpes erythrocephalus (*Linn.*). Redheaded Woodpecker.

A rare summer resident of some parts of the state, more common in the autumn migration than at any other season, and even then it is quite rare.

County Records.—Androscoggin, (Pike); Cumberland, "rare, irregular transient" (Brown's Cat. Birds of Portland, p. 19); Franklin, "rare" (Richards); Kennebec, (Larrabee); Knox, "occasional visitant" (Rackliff); Oxford, "visitant" (Nash); Penobscot, "very rare, has been taken in July" (Knight); Piscataquis, "rare" (Homer); Sagadahoc, "rare, only three specimens, all in fall" (Spinney); Washington, "very rare" (Boardman); York, "breeds sparingly" (Adams).

Genus COLAPTES Swainson.

183. (412). Colaptes auratus (*Linn.*). Flicker.

The "Yellow-hammer" is the most abundant of our Woodpeckers throughout the state, where it is a summer resident, occurring from the last of April to late October. They are especially fond of a large species of black ant which occurs here, and may often be seen on the ground near an ant hill, feeding upon the occupants thereof. Cherries and other small fruits are also welcome additions to their bill of fare.

County Records.—Androscoggin, "abundant summer resident" (Johnson); Aroostook, "rather common at Fort Fairfield" (Batchelder, Bull. Nutt. Orn. Club, Vol. 7, p. 150); Cumberland, "common summer resident" (Mead); Franklin, "common summer resident" (Swain); Hancock, "common summer resident" (Murch), "I have found this bird to be very common in summer on the various wooded islands of the coast" (Knight); Kennebec, "quite common summer resident" (Gardiner Branch); Knox, "summer resident" (Racklif); Oxford, "common, breeds" (Nash); Penobscot, "breeds commonly and is the most abundant of our woodpeckers" (Knight); Piscataquis, "common" (Homer); Sagadahoc, "common summer resident" (Spinney); Somerset, "common summer resident" (Morrell); Waldo, "common summer resident" (Knight); Washington, "abundant summer resident" (Boardman); York, "not common, formerly abundant" (Adams).

Order MACROCHIRES. Goatsuckers, Swifts, etc.

Suborder CAPRIMULGI. Goatsuckers, etc.

Family CAPRIMULGIDÆ. Goatsuckers, etc.

Genus ANTROSTOMUS Gould.

184. (417). Antrostomus vociferus (Wils.). Whip-poor-will.

A fairly common summer resident throughout the state and in some parts it is very abundant. On account of its retiring habits during the day time it is a bird more often heard than seen, but its cry of "whip-poor-will, whip-poor-will," which one may hear on the quiet evenings of late May and June, is a very good proof of its presence in any locality.

County Records.—Androscoggin, "common summer resident" (Johnson); Aroostook, "found at Houlton" (Batchelder, Bull. Nutt. Orn. Club, Vol. 7, p. 150); Cumberland, "common summer resident" (Mead); Franklin, "common summer resident" (Richards); Hancock, "summer resident" (Dorr); Kennebec, "rare summer resident" (Gardiner Branch);

Knox, "migrant" (Rackliff); Oxford, "breeds commonly" (Nash); Penobscot, "common near the Maine State College, where on calm June evenings I have heard six or eight of these birds calling as I sat studying, and I have often been awakened during the night by one holding forth on the roof of the house over my head" (Knight); Piscataquis, "common" (Homer); Sagadahoc, "very rare" (Spinney); Somerset, "not common summer resident" (Morrell); Waldo, (Spratt); Washington, "not uncommon summer resident" (Boardman); York, "common summer resident" (Adams).

Genus CHORDEILES Swainson.

185. (420). Chordeiles virginianus (*Gmel.*). Nighthawk.

A common summer resident, breeding throughout the entire state. It arrives from the south about the middle or last of May and departs in late August, usually migrating in large bands. A few stragglers are found up to the middle of September.

County Records.—Androscoggin, "common summer resident" (Johnson); Aroostook, "common at Fort Fairfield" (Batchelder, Bull. Nutt. Orn. Club, Vol. 7, p. 150); Cumberland, "common summer resident" (Mead); Franklin, "common summer resident" (Swain); Hancock, "summer resident" (Murch); Kennebec, "quite common summer resident" (Gardiner Branch); Knox, "summer resident" (Rackliff); Oxford, "breeds commonly" (Nash); Penobscot, "common summer resident, I have found it breeding on flat gravelled roofs of buildings in the heart of the city of Bangor" (Knight); Piscataquis, "common" (Homer); Sagadahoc, "common summer resident" (Spinney); Somerset, "common summer resident" (Morrell); Waldo, (Spratt); Washington, "abundant summer resident" (Boardman); York, "common summer resident" (Adams).

Suborder CYPSELI.   Swifts.

Family MICROPODIDÆ.   Swifts.

Subfamily CHÆTURINÆ.   Spine-tailed Swifts.

Genus CHÆTURA Stephens.

186. (423). Chætura pelagica (*Linn.*).   Chimney Swift.

A common summer resident throughout the state, usually placing its nests in the disused chimneys of some house. Near Bucksport there is a large, disused chimney of a storehouse where fully 100 of these birds make their home in summer, as I am informed by Mr. Dorr of that town. Usually only one or two pair of birds are found inhabiting a single chimney, but I have personally seen one containing ten nests with eggs, and see no reason why they should not be more numerous in a given chimney under favorable circumstances. This species also attaches its nests to the inner walls of barns and other buildings when no chimney is convenient, and I suspect that in isolated parts of the state the ancient custom, in vogue before the advent of civilization, of placing the nest against the inner wall of a hollow tree, is still adhered to.

County Records.—Androscoggin, "abundant summer resident" (Johnson); Aroostook, "seen at Fort Fairfield" (Batchelder, Bull. Nutt. Orn. Club, Vol. 7, p. 150); Cumberland, "common summer resident" (Mead); Franklin, "common summer resident" (Swain); Hancock, "summer resident" (Murch); Kennebec, "abundant summer resident" (Gardiner Branch); Knox, "summer resident" (Rackliff); Oxford, "breeds commonly" (Nash); Penobscot, "abundant in the settled parts of the county" (Knight); Piscataquis, "common summer resident" (Whitman); Sagadahoc, "common summer resident" (Spinney); Somerset, "common summer resident" (Morrell); Waldo, "common summer resident" (Knight); Washington, "abundant summer resident" (Boardman); York, "common summer resident" (Adams).

Suborder TROCHILI. Hummingbirds.

Family TROCHILIDÆ. Hummingbirds.

Genus TROCHILUS Linnæus.

Subgenus TROCHILUS.

187. (428). Trochilus colubris *Linn.* Ruby-throated Hummingbird.

Common summer resident everywhere through the state, but perhaps occurring in slightly greater abundance in the southern part.

County Records.—Androscoggin, "fairly common summer resident" (Johnson); Aroostook, "seen at Fort Fairfield" (Batchelder, Bull. Nutt. Orn. Club, Vol. 7, p. 150); Cumberland, "common summer resident" (Mead); Franklin, "common summer resident" (Swain); Hancock, "summer resident" (Murch); Kennebec, "common summer resident" (Gardiner Branch); Knox, "summer resident" (Rackliff); Oxford, "breeds commonly" (Nash); Penobscot, "fairly common summer resident" (Knight); Piscataquis, "common" (Homer); Sagadahoc, "common summer resident" (Spinney); Somerset, "quite common summer resident" (Morrell); Waldo, (Spratt); Washington, "abundant summer resident" (Boardman); York, "summer resident" (Adams).

Order PASSERES.   Perching birds.

Suborder CLAMATORES.   Songless Perching Birds.

Family TYRANNIDÆ.   Tyrant Flycatchers.

Genus TYRANNUS Cuvier.

188.   (444).   Tyrannus tyrannus (*Linn.*).   Kingbird.

Everywhere a common summer resident, but nevertheless this
species prefers to make its summer home in the immediate vicinity
of some dwelling house when possible.   The nest is often placed in
an apple tree, although where I found this bird nesting away from
the vicinity of houses it often placed its domicile in some elm or
maple.   They also nest in dead trees overhanging the water.
It is regarded with disfavor by bee keepers on account of its liking
for bees, but probably a large part of these insects which fall vic-
tims to its appetite are drones and consequently of no value to the
apiarist.

County Records.—Androscoggin, "common summer resident" (Johnson); Aroostook, "seen at Fort Fairfield" (Batchelder, Bull. Nutt. Orn.
Club, Vol. 7, p. 149); Cumberland, "common summer resident" (Mead);
Franklin, "common summer resident" (Swain); Hancock, "summer
resident" (Murch); Kennebec, "abundant summer resident" (Gardiner
Branch); Knox, "summer resident" (Rackliff); Oxford, "common sum-
mer resident" (Johnson); Penobscot, "breeds commonly" (Knight);
Piscataquis, "common" (Homer); Sagadahoc, "common summer resi-
dent" (Spinney); Somerset, "common summer resident" (Morrell);
Waldo, "common summer resident" (Knight); Washington, "very
abundant" (Boardman); York, "quite common summer resident"
(Adams).

189.   (447).   Tyrannus verticalis *Say*.   Arkansas Kingbird.

A single specimen has been taken at Elliot, York County, so this
species is entitled to a place in our fauna as a purely accidental
visitor.   (Cf. Purdie, Bulletin of the Nuttall Ornithological Club,
Vol. 1, p. 73).

Genus MYIARCHUS Cabanis.

190.   (452).   Myiarchus crinitus (*Linn.*).   Crested Flycatcher.

A summer resident in the southern part of the state, and uncommon within the limits of the Canadian fauna.

County Records.—Androscoggin, "fairly common summer resident" (Johnson); Cumberland, "uncommon summer resident" (Brown's Cat. Birds of Portland, p. 17); Franklin, "rare summer resident" (Swain); Kennebec, "summer resident" (Dill); Oxford, "breeds rarely" (Nash); Penobscot, "rare summer resident, usually only two or three individuals observed in the course of a season" (Knight); Piscataquis, "not uncommon, breeds" (Homer); Somerset, "quite common summer resident" (Morrell); Waldo, (Spratt); Washington, "very rare" (Boardman); York, "rare summer resident "(Adams).

Genus SAYORNIS Bonaparte.

191. (456). Sayornis phœbe (*Lath.*). Phœbe.

A common summer resident which is locally known as "Bridge Pewee" from its propensity for placing its nest under bridges, where the beams overhead serve as a foundation for the domicile. The names Phœbe and Pewee are given the bird on account of its notes which resemble these words.

County Records.—Androscoggin, "common summer resident" (Johnson); Aroostook, "seen at Fort Fairfield and Houlton" (Batchelder, Bull. Nutt. Orn. Club, Vol. 7, p. 149); Cumberland, "common summer resident" (Mead); Franklin, "common summer resident" (Swain); Hancock, "summer resident" (Murch); Kennebec, "common summer resident" (Gardiner Branch); Knox, "visitant" (Norton); Oxford, "breeds commonly" (Nash); Penobscot, "common summer resident" (Knight); Piscataquis, "common, breeds" (Homer); Sagadahoc, "common summer resident" (Spinney); Somerset, "common summer resident" (Morrell); Waldo, (Spratt); Washington, "rare" (Boardman); York, "not common summer resident" (Adams).

Genus CONTOPUS Cabanis.
Subgenus NUTTALLORNIS Ridgw.

192. (459). Contopus borealis (*Swains.*). Olive-sided Flycatcher.

A summer resident chiefly confined to the Canadian fauna, elsewhere it occurs quite commonly in the migrations.

County Records.—Androscoggin, "has been seen here" (Walters, The Birds of Androscoggin Co., p. 23); Aroostook, "rather common at Fort Fairfield" (Batchelder, Bull. Nutt. Orn. Club, Vol. 7, p. 149); Cumberland, "rare" (Mead); Franklin, "rare summer resident" (Swain); Kennebec, "rare summer resident" (Dill); Knox, "summer" (Rackliff); Oxford, "breeds rarely" (Nash); Penobscot, "tolerably common summer resident" (Knight); Piscataquis, "common, breeds" (Homer); Washington, "not uncommon summer resident" (Boardman).

Subgenus CONTOPUS Cabanis.

193. (461). Contopus virens (*Linn.*). Wood Pewee.

Common summer resident throughout the state, frequenting the trees of both woodland and city. The characteristic, drawling note "pe-wee-a-a-" is one of the best evidences of its presence, even when the bird is not seen.

County Records.—Androscoggin, "fairly common summer resident" (Johnson); Aroostook, "not uncommon at Fort Fairfield" (Batchelder, Bull. Nutt. Orn. Club, Vol. 7, p. 149); Cumberland, "common summer resident" (Mead); Franklin, "common summer resident" (Swain); Hancock, "summer resident" (Knight); Kennebec, "quite common summer resident" (Gardiner Branch); Knox, "rare in summer" (Racklift); Oxford, "common, breeds" (Nash); Penobscot, "common throughout the summer" (Knight); Piscataquis, "common summer resident" (Homer); Sagadahoc, "summer resident" (Spinney); Somerset, "common summer resident" (Morrell); Waldo, (Spratt); Washington, "not uncommon summer resident" (Boardman).

Genus EMPIDONAX Cabanis.

194. (463). Empidonax flaviventris *Baird.* Yellow-bellied Flycatcher.

A somewhat rare summer resident of those counties within the Canadian fauna, while in other parts of the state it occurs as a rare migrant.

County Records.—Androscoggin, (Walter's Birds of Androscoggin County, p. 23); Aroostook, "rather common at Fort Fairfield, breeds" (Batchelder, Bull. Nutt. Orn. Club, Vol. 7, p. 149); Cumberland, "quite rare transient" (Brown's Cat. Birds of Portland, p. 18); Franklin, "rare summer resident" (Richards); Hancock, "rare summer resident" (Knight); Kennebec, "rare" (Gardiner Branch); Knox, "summer resident" (Norton); Oxford, "breeds at Richardson Lake" (Cf. Osborne, Bull. Nutt Orn. Club, Vol. 4, p. 240); Penobscot, "quite common summer resident" (Knight); Somerset, "not common migrant" (Morrell); Waldo, (Spratt); Washington, "not uncommon summer resident" (Boardman).

195. (466a). Empidonax traillii alnorum *Brewst.* Alder Flycatcher.

A fairly common summer resident in most parts of the state, but owing to its frequenting alder thickets and other low bushy areas and being rarely met with outside of its favorite habitat, this species has not been reported from many localities where it undoubtedly occurs. This is the Traill's Flycatcher of previous lists, but as pointed out by Mr. Brewster in The Auk, Vol. 12, p. 161, Audubon's type of *traillii* was from the west, and our eastern bird being subspecifically distinct, this name is given it by him.

County Records.—Androscoggin, "fairly common summer resident" (Johnson); Aroostook, "seen at Houlton" (Purdie, Bull. Nutt. Orn. Club, Vol. 1, p. 76); Cumberland, "common summer resident" (Brown's Cat. Birds of Portland, p. 18); Franklin, "common summer resident" (Swain); Hancock, "quite common summer resident" (Knight); Kennebec, "rare summer resident" (Robbins); Knox, "summer" (Racklift); Oxford, "rare breeder" (Nash); Penobscot, "quite a common summer resident, but not met with except in alder thickets" (Knight); Piscataquis, "common summer resident" (Whitman); Sagadahoc, "rare summer resident" (Spinney); Somerset, "common summer resident" (Morrell); Waldo, (Knight); Washington, "not uncommon summer resident" (Boardman).

196. (467). Empidonax minimus *Baird.* Least Flycatcher.

Common summer resident throughout the state. This species can always be distinguished from all the other Flycatchers found here by its cry of "ché-bèc, ché-bèc." The bird in hand when compared with the Alder Flycatcher resembles it very closely, but the wing of the latter is over 2.60 inches in length while the wing of the Least Flycatcher is under 2.60, and there is also a slight difference in the color of the wing coverts. The notes of the two species are altogether different, as are the nests and eggs, and the habitats of the birds, the Least Flycatcher always frequenting open woods, orchards, or the trees of our towns and cities, while *alnorum* is always confined to alder growths.

County Records.—Androscoggin, "common summer resident" (Johnson); Aroostook, "rather common at Fort Fairfield" (Batchelder, Bull. Nutt. Orn. Club, Vol. 7, p. 149); Cumberland, "common summer resident" (Mead); Franklin, "common summer resident" (Swain); Hancock, "summer resident" (Dorr); Kennebec, "common summer resident" (Gardiner Branch); Oxford, "common breeder" (Nash); Penobscot, "very common summer resident" (Knight); Piscataquis, "common, breeds" (Homer); Sagadahoc, "common" (Spinney); Somerset, "common summer resident" (Morrell); Waldo, (Spratt); Washington, "abundant summer resident" (Boardman); York, "common summer resident" (Adams).

Suborder OSCINES.   Song Birds.

Family ALAUDIDÆ.   Larks.

Genus OTOCORIS Bonaparte.

197.   (474).   Otocoris alpestris (*Linn.*).   Horned Lark.

Occurs only as a winter resident and then is subject to periods of variable abundance. It is of somewhat local occurrence, being met with commonly in some places, while not occurring at all in others.

County Records.—Androscoggin, "rare migrant" (Johnson); Cumberland, "gregarious winter resident of variable abundance" (Brown's Cat. Birds of Portland, p. 17); Knox, "winter" (Rackliff); Oxford, "occurs in winter at Norway" (Verrill's List of Birds of Norway); Piscataquis, "common" (Homer); Washington, "very rare" (Boardman).

198.   (474b).   Otocoris alpestris praticola *Hensh.*   Prairie Horned Lark.

The published records of this subspecies and the preceding species have been inextricably mixed, and consequently I have been obliged to ignore all previous records of either and only admit records which are made on the basis of specimens actually in existence. The first published record of the occurrence of *praticola* in Maine will be found in the Maine Sportsman for April, 1897, page 6. Mr. J. C. Mead of North Bridgton, the author of the above cited article writes me as follows : "On March 13th, 1897, my attention was called to a flock of about 25 birds feeding busily in the street near the outskirts of our village (North Bridgton, Cumberland Co.). I secured four specimens. Measurements and descriptions led me to believe that I had secured the Prairie Horned Lark. Mr. A. H. Norton who has examined all the skins pronounces them typical of this variety." Mr. Norton writes that two of these are males by dissection, one a male by proportions and markings, and the remaining one a female by dissection. A pair of these birds in the collection of the University of Maine are typical specimens. They were killed at Bucksport, Hancock Co. about '86 or '87, in the winter, by Alvah G. Dorr. Mr. Harry Merrill has a typical male which was killed at Bangor, Penobscot Co., March 30th, 1887. Mr. Wallace Homer of Monson, Piscataquis Co., has a fairly typical example which is a male according to appearances. Prof. A. L. Lane of Waterville, Kennebec County, has a specimen, seemingly a male, which was taken at that city in the spring of '92 or '93. Mr. C. H. Morrell of Pittsfield, Somerset County, has a male taken there March 29th, 1892, one taken March 27th, 1893, and a female taken March 22nd, 1894. Mr. C. D. Farrar of Lewiston, Androscoggin County, took a specimen from a flock of 8 or 10, February 26th, 1897. Part of these have been verified by Mr. Brewster, and the identification of the remainder rests upon the author. With the foregoing evidence we may safely say that the Prairie Horned Lark is a regular migrant in many parts of the state, and it is not improbable that it may ultimately be found breeding within our limits.

Family CORVIDÆ. Crows, Jays, Magpies, etc.

Subfamily GARRULINÆ. Magpies and Jays.

Genus CYANOCITTA Strickland.

199. (477). Cyanocitta cristata (*Linn.*). Blue Jay.

A tolerably common resident of general distribution, usually frequenting dense woods in this state, and rarely venturing within the limits of our villages and towns. In Illinois, Kansas, and other western states this bird is quite different in its habits, venturing freely and boldly into the towns and villages, and constructing its nest in trees in the very dooryards. I was much surprised to find our Maine Blue Jays so shy and unsociable, being quite the opposite of the tame, fearless Jays I had known in the west.

County Records.—Androscoggin, "common resident" (Johnson); Aroostook, "rather common at Fort Fairfield" (Batchelder, Bull. Nutt. Orn. Club, Vol. 7, p. 149); Cumberland, "common resident" (Mead); Franklin, "common resident" (Swain); Hancock, "common resident"; (Dorr); Kennebec, "quite common resident" (Sanborn); Knox, "summer" (Racklift); Oxford, "breeds commonly" (Nash); Penobscot, "resident, commonest in fall, in summer this species retires to the most isolated localities to nest and in such places it is not uncommon" (Knight); Piscataquis, "common resident" (Homer); Sagadahoc, "common resident" (Spinney); Somerset, "not very common resident" (Morrell); Waldo, (Spratt); Washington, "common resident" (Boardman); York, "all too common" (Adams).

Genus PERISOREUS Bonaparte.

200. (484). Perisoreus canadensis (*Linn.*). Canada Jay.

A typical bird of the Canadian fauna and resident within its limits. Elsewhere in the state it occurs only as a straggler. About

the lumber camps of northern Maine it is one of the commonest
and most familiar of birds, and here it is known as the Moose Bird
or Whiskey Jack.

County Records.—Androscoggin, "rare visitant" (Johnson); Aroostook, "very common at Houlton" (Batchelder, Bull. Nutt. Orn. Club,
Vol. 7, p. 149); Cumberland, "rare" (Mead); Franklin, "very rare"
(Swain); Hancock, "rare" (Dorr); Knox, "rare migrant" (Racklift);
Oxford, "breeds rarely" (Nash); Penobscot, "common in northern part
of the county, quite rare in the southern part" (Knight); Piscataquis,
"common resident" (Homer); Somerset, "found near Jackman" (Harvey
and Knight); Washington, "common resident" (Boardman); York, "a
few seen" (Adams).

Subfamily CORVINÆ.   Crows.

Genus CORVUS Linnæus·

201. (486a). Corvus corax principalis *Ridgw.*   Northern
Raven.

A quite common resident along the coast and of rare occurrence
in the interior.   On May 16, 1896, I observed a nest of this species containing nearly fledged young, on an island in Penobscot
Bay.   A colony of Black-crowned Night Herons were breeding on
the same island, but no eggs were found in any of the nests save
two.   Under the nests were dozens of the eggs with "bill holes" in
them which clearly demonstrated the use they had been put to by
the Ravens.   In June I again visited the island and found the
young Ravens flying about, while the Herons had been driven from
the place by the constant prosecutions they had been subjected to.

County Records.—Aroostook, "rare at Houlton" (Batchelder, Bull.
Nutt. Orn. Club, Vol. 7, p. 149); Cumberland, "very rare winter visitor"
(Brown's Cat. Birds of Portland, p. 17); Franklin, "rare or accidental"
(Richards); Hancock, "resident and nests on many of the islands along
the coast" (Knight); Knox, "resident" (Racklift); Lincoln, "seen in
June, 1897" (Norton); Oxford, "visitant" (Nash); Penobscot, (Hardy);
Sagadahoc, "common, nests" (Spinney); Washington, "not common,
breeds" (Boardman).

202. (488.). Corvus americanus *Aud.*   American Crow.

A common resident along the coast, a common summer resident
throughout the state, and of rare occurrence in winter in the interior.   The Crow is a much maligned bird and undoubtedly does far
more good than harm.   The small amounts of corn and grain

which are eaten are more than paid for by the great numbers of
injurious insects which are devoured. Along the coast Crows may
be seen at low tide, feeding on the various forms of marine animals
which are exposed by the receding waters. I have very good evidence that this species also visits the outer islands and feeds on the
eggs of the Black Guillemots and Terns. I have seen them leave
islands where these birds nested, and on landing found fresh frag-
ments of partly devoured eggs with "bill holes" in them. It is a
mystery how they manage to get the eggs of the Guillemots,
as they are always deposited far under piles of rocks, and the
ingenuity of a collector is taxed to find them. On the whole I
believe the Crow is rather more beneficial than injurious to the
farmer.

County Record.—Androscoggin, "abundant summer resident" (John-
son); Aroostook, "common at Fort Fairfield" (Batchelder, Bull. Nutt.
Orn. Club, Vol. 7, p. 149); Cumberland, "common resident" (Mead);
Franklin, "common resident" (Swain); Hancock. "common resident"
(Knight); Kennebec, "quite common summer resident, rarely resident"
(Gardiner Branch); Knox, "resident" (Rackliff); Lincoln, "common"
(Norton); Oxford, "common, breeds" (Nash); Penobscot, "abundant
summer resident, rare in winter" (Knight); Piscataquis, "summer resident" (Homer); Sagadahoc, "common" (Spinney); Somerset, "common
summer resident" (Morrell); Waldo, "common resident along the coast,
probably common summer resident in interior" (Knight); Washington,
"common resident" (Boardman); York, "common" (Adams).

Family STURNIDÆ. Starlings.
Genus STURNUS Linnæus.

203. (493). Sturnus vulgaris *Linn.* Starling.

Mr. George A. Boardman informs me that it is accidental at
Calais, a specimen having been shot by a Mr. Nichols, on May 4,
1889.

Family ICTERIDÆ. Blackbirds, Orioles, etc.
Genus DOLICHONYX Swainson.

204. (494). Dolichonyx oryzivorus (*Linn.*). Bobolink.

A common summer resident wherever there are grassy meadows
and fields throughout the state. The male is well known by
his wild, happy, rollicking song. The dull-colored female is
less apt to be noticed, and fewer people are acquainted with

the quaker wife of one of our best songsters. In fall the males take on a dull-colored plumage and their sole cry becomes a low chirp or chink. In early September they leave for the rice marshes of the south and here they are known as Rice Birds. In New York restaurants they are served to epicures under the name of Reed Birds, although the despised English Sparrow is now being brought into their place under the same name.

County Records.—Androscoggin, "common summer resident" (Johnson); Aroostook, "not rare at Fort Fairfield" (Batchelder, Bull. Nutt. Orn. Club, Vol. 7, p. 149) Cumberland, "common summer resident" (Mead); Franklin, "common summer resident" (Swain); Hancock, "summer resident" (Murch); Kennebec, "common summer resident" (Gardiner Branch); Knox, "summer" (Racklift); Oxford, "common, breeds" (Nash); Penobscot, "very common breeder" (Knight); Piscataquis, "common, breeds" (Homer); Sagadahoc, "common summer resident" (Spinney); Somerset, "common summer resident" (Morrell); Waldo, (Spratt); Washington, "very abundant summer resident" (Boardman); York, "rarely breeds" (Adams).

Genus MOLOTHRUS Swainson.

205. (495). Molothrus ater (*Bodd.*). Cowbird.

A common summer resident in most parts of the state and of very general distribution. This species makes no nest of its own, but instead deposits its eggs in the nests of smaller birds, usually laying only one egg in a nest though as many as four eggs of the Cowbird have been found in one nest, these probably being the product of as many different females.. The nests of our Warblers are perhaps thus imposed on most often, though the Chipping Sparrow, Red-eyed Vireo, and other small birds are also forced to receive the intruding egg into their nests.

County Records.—Androscoggin, "common summer resident" (Johnson); Cumberland, "common summer resident" (Mead); Franklin, "common summer resident" (Swain); Hancock, "summer resident" (Dorr); Kennebec, "quite common summer resident" (Gardiner Branch); Knox, "migrant" (Racklift); Oxford, "breeds commonly" (Nash); Penobscot, "a common summer resident" (Knight); Piscataquis, "not common" (Homer); Sagadahoc, "common summer resident" (Spinney); Somerset, "common summer resident" (Morrell); Waldo, (Spratt); Washington, "rare summer resident" (Boardman).

Genus XANTHOCEPHALUS Bonaparte.

206. (497). Xanthocephalus xanthocephalus (*Bonap.*). Yellow-headed Blackbird.

Of purely accidental occurrence, only one specimen having been taken in the state. This was taken by Mr. Rackliff, at Spruce Head, Knox County, on August 17, 1882. (Cf. Norton, The Auk, Vol. 11, pp. 78-79, and also in the same connection Cf. Ridgway, The Auk, Vol. 4, p. 256; the notes by Mr. Norton give more details of this bird and also corrections regarding the date of capture).

Genus AGELAIUS Vieillot.

207. (498). Agelaius phœniceus (*Linn.*). Red-winged Blackbird.

Common everywhere in migrations, while in summer it is locally abundant wherever marshes and cat-tail swamps furnish an abiding place.

County Records.—Androscoggin, "abundant summer resident" (Johnson); Aroostook, "quite common on Eel River" (Batchelder, Bull. Nutt. Orn. Club, Vol. 7, p. 149); Cumberland, "common summer resident" (Mead); Franklin, "common summer resident" (Swain); Hancock, "breeds" (Murch); Kennebec, "common summer resident" (Gardiner Branch); Knox, "summer"(Rackliff); Lincoln, (Norton); Oxford,"breeds commonly" (Nash); Penobscot, "common summer resident in suitable localities"(Knight); Piscataquis,"common,breeds" (Homer); Sagadahoc, "common summer resident"(Spinney); Somerset, "common summer resident"(Morrell); Washington, "abundant summer resident" (Boardman); York, "common summer resident (Adams).

Genus STURNELLA Vieillot.

208. (501). Sturnella magna (*Linn.*). Meadow Lark.

A rare summer resident of local occurrence. Its scarcity in parts of the state is partly due to the lack of suitable expanses of meadow and grass land which this species loves to frequent, and partly to its being a typical Alleghanian species.

County Records.—Androscoggin, "rare migrant" (Johnson); Cumberland, "rare summer resident, oftenest seen in migration" (Brown's Cat. Birds of Portland, p. 16); Franklin, "rare summer resident" (Swain); Kennebec, (Larrabee); Knox, "rare migrant" (Rackliff); Oxford, "breeds rarely" (Nash); Penobscot, "a pair of these birds frequented the same field in the summers of 1894 and 1895, and their nest was found by some small boys the first season; outside of this pair of birds I have never seen more than three other individuals in the county" (Knight); Piscataquis, "rare" (Homer); Sagadahoc, "rare, one specimen" (Spinney); Somerset, "rare summer resident" (Morrell); Washington, "accidental" (Boardman).

Genus ICTERUS Brisson.

Subgenus PENDULINUS Vieillot.

209. (506). Icterus spurius (*Linn.*). Orchard Oriole.

Accidental within the state, only three instances of its occurrence being known, these all based upon the capture of specimens.

County Records.—Androscoggin, "have one taken near Auburn" (Pike); Knox, "a specimen was taken at Thomaston by Chas. A. Creighton" (E. Smith); Washington, "accidental, a male taken here in the sixties" (Boardman).

Subgenus YPHANTES Vieillot.

210. (507). Icterus galbula (*Linn.*). Baltimore Oriole.

A summer resident of the southern and western part of the state, where it is quite common, while in the eastern and northern parts it occurs only as a straggler.

County Records.—Androscoggin, "common summer resident" (Johnson); Cumberland, "common summer resident" (Mead); Franklin, "common summer resident" (Swain); Hancock, "summer resident" (Murch); Kennebec, "quite common summer resident" (Gardiner Branch); Knox, "summer" (Rackliff); Oxford, "breeds commonly" (Nash); Penobscot, "common summer resident of the southern part of the county while in the northern part it does not occur to my knowledge" (Knight); Piscataquis, "not common, breeds" (Homer); Sagadahoc, "rare, three specimens" (Spinney); Somerset, "common summer resident" (Morrell); Waldo, (Spratt); Washington, "straggler" (Boardman); York, "quite common summer resident" (Adams).

211. (508). Icterus bullocki (Swains.). Bullock's Oriole.

This western bird is entitled to a place in the list upon the strength of a specimen taken at Sorrento, Hancock County, and now in the collection of Mr. Manly Hardy of Brewer. What is undoubtedly this same specimen is recorded by Mr. Brewster in The Auk, Vol. 7, p. 92, although he erroneously gives the locality as "near Bangor."

Genus SCOLECOPHAGUS Swainson.

212. (509). Scolecophagus carolinus (*Müll.*). Rusty Blackbird.

A common migrant of general occurrence, while in the extreme northern counties of the Canadian fauna it breeds to some extent. It has been reported as nesting in the Magalloway region.

County Records.—Androscoggin, "fairly common migrant" (Johnson); Cumberland, "migrant" (Mead); Franklin, "rare summer resident, specimens shot late in June" (Swain); Kennebec, (Larrabee); Knox, "migrant" (Racklift); Oxford, (Nash); Penobscot, "common migrant" (Knight); Piscataquis, "common summer resident" (Homer); Sagadahoc, "common migrant" (Spinney); Somerset, "common migrant" (Morrell); Waldo, (Spratt); Washington, "common migrant and rare summer resident" (Boardman).

Genus QUISCALUS Vieillot.
Subgenus QUISCALUS.

213. (511b). Quiscalus quiscula æneus (*Ridgw*.). Bronzed Grackle.

This is the Purple Grackle of previous lists, although it is well to emphasize the point that the true Purple Grackle has never been taken in the state. Previous to the pointing out of the distinctive characteristics of the Bronzed Grackle, all these birds were included under the above named species, and for this reason it has been given in previous lists as Purple Grackle.

County Records.—Androscoggin, "fairly common summer resident" (Johnson); Aroostook, "common at Fort Fairfield" (Batchelder, Bull. Nutt. Orn. Club, Vol. 7, p. 149); Cumberland, "common summer resident" (Mead); Franklin, "common summer resident" (Swain); Hancock, "summer resident" (Murch); Kennebec, "rare summer resident" (Gardiner Branch); Knox, "migrant" (Rackliff); Oxford, "breeds rarely" (Nash); Penobscot, "common summer resident" (Knight); Piscataquis, "common summer resident" (Homer); Sagadahoc, "rare, three specimens" (Spinney); Somerset, "common summer resident" (Morrell); Washington, "very abundant summer resident" (Boardman); York, (Adams).

Family FRINGILLIDÆ. Finches, Sparrows, etc.
Genus COCCOTHRAUSTES Brisson.
Subgenus HESPERIPHONA Bonaparte.

214. (514). Coccothraustes vespertinus (*Coop*.). Evening Grosbeak.

A casual visitor from the west which was recorded from a number of the eastern states during the winter of 1889-90, when an extensive eastern movement of these birds took place.

County Records.—Androscoggin, "a male taken on the Bates College campus, January 10th, 1890" (Walter, The Birds of Androscoggin County, p. 14); Oxford, "I mounted one taken at Fryeburg" (Nash);

Penobscot, "Mr. S. L. Crosby has seen a specimen of this bird, shot in Brewer in the winter of 1889-90, while Geo. P. Shepherd records another specimen from Bangor, taken March 18, 1890, while a companion which it had escaped." (For this last Cf. Shepherd, The Oologist, May 1890, p. 86).

Genus PINICOLA Vieillot.

215. (515). Pinicola enucleator (*Linn.*). Pine Grosbeak.

This is a fairly regular winter visitor of varying abundance. n/ During the winters of 1895-96 and 1896-97 it was very abundant. / In 1893-94 none of these birds were seen near Bangor, and they were seemingly equally rare throughout the state. In captivity these birds make interesting and affectionate pets, and the song (both sexes sing though the males sing more frequently) is, according to my opinion, far sweeter in tone than that of the Canary. It is interesting to note that while in nature the adult males are adorned with carmine crown, breast, back, and upper tail coverts, this color changes to a pale orange at the first moult which takes place in captivity. Young males, which are like the females in general appearance during the first year, also take on this orange color in captivity, instead of assuming the carmine garb when they reach maturity. For an account of the habits of this bird in captivity see Knight, The Auk, Vol. 13, pp. 21-24. It is a rare summer resident in the northern and eastern counties of the Canadian fauna.

County Records.—Androscoggin, "common winter visitant" (Johnson); Cumberland, "common winter migrant" (Mead); Franklin, "common winter resident" (Swain); Hancock, "irregular winter migrant" (Murch); Kennebec, "common winter visitor" (Gardiner Branch); Knox, "winter" (Rackliff); Oxford, "common in winter, rare in summer" (Nash); Penobscot, "usually common in winter, and especially so for the last two seasons, '96-'97" (Knight); Piscataquis, "common winter visitor" (Homer); Sagadahoc, "irregularly common winter visitor" (Spinney); Somerset, "irregularly common winter resident" (Morrell); Waldo, (Spratt); Washington, "common in winter, rare summer resident" (Boardman); York, "regular winter visitant" (Adams).

Genus CARPODACUS Kaup.

216. (517). Carpodacus purpureus (*Gmel.*). Purple Finch.

A common summer and rare winter resident in most parts of the state. While this species seems to prefer to frequent the neigh-

borhood of houses in the settled parts of the state, yet in the back-
woods I have seen it far from any dwelling, and in as great
abundance as it occurs elsewhere.

County Records.—Androscoggin, "common summer resident" (Johnson) Aroostook, "common at Fort Fairfield" (Batchelder, Bull. Nutt.
Orn. Club, Vol. 7, p. 147); Cumberland, "common summer resident"
(Mead), "a few of these birds wintered about Westbrook and Gorham
through '91-'92" (Norton);Franklin,"common summer resident" (Swain);
Hancock, "summer resident" (Murch); Kennebec, "rare resident" (Gardiner Branch); Knox, "summer" (Rackliff); Oxford, "breeds commonly" (Nash); Penobscot, "common summer and rare winter resident"
(Knight); Piscataquis, "breeds, resident in mild winters" (Homer);
Sagadahoc, "common except in midwinter" (Spinney); Somerset, "quite
common summer resident" (Morrell); Waldo, (Spratt); Washington,
"abundant, breeds" (Boardman); York, "common migrant, may breed"
(Adams).

Genus LOXIA Linnæus.

217. (521). Loxia curvirostra minor (*Brehm*). American
Crossbill.

A resident species but not found in any one place in
numbers through the season. The Crossbills are among the
most irregular and eccentric of birds, breeding at almost
any season of the year and in any part of the state where the
impulse to do so comes to them. They have been reported breeding in February in other states. At Jackman in the latter part of
August, 1895, Prof. F. L. Harvey and myself observed old birds
feeding their young, evidently not long from the nest, and also saw
paired birds flying about. (Cf. Knight, The Auk, Vol. 12, pp.
390-91). Mr. Manly Hardy of Brewer also informs me that his
son found a nest in June, some years ago, a short distance back of
his residence in the above named city. It was situated in a juniper
tree and was taken with the female parent. The eggs were unfortunately broken.

County Records.—Androscoggin, "fairly common winter visitor"
(Johnson); Cumberland, "common winter migrant" (Mead); Franklin,
"common resident" (Richards); Hancock, "common in winter and I
have observed it on Pickering's Island in May and also June" (Knight);
Kennebec, "very rare resident" (Powers); Knox, "winter" (Rackliff);
Oxford, "breeds" (Nash); Penobscot, "irregularly abundant, I have
seen the species every month of the year" (Knight); Piscataquis, "common in winter" (Homer); Sagadahoc, "irregularly common winter visi-

tor" (Spinney); Somerset, "irregular winter visitant" (Morrell); Waldo,
"seen in May" (Knight); Washington, "uncertain, some winters abundant, breeds in winter" (Boardman); York, "migrant" (Adams).

218. (522). Loxia leucoptera *Gmel.* White-winged Crossbill.

A resident species of much rarer occurrence than the preceding
and more liable to occur in winter. It seemingly nests exclusively
in winter, as it has not been reported breeding in summer like the
American Crossbill. Much still remains unknown regarding the
life histories of our Crossbills. The two species often occur in the
same flock though *leucoptera* always occurs least abundantly. Said
to be common all the year, about the lumber camps with the preceding.

County Records.—Androscoggin, "rare winter visitor" (Johnson);
Cumberland, "rare winter migrant" (Mead); Franklin, "common winter
resident" (Richards); Kennebec, "very rare resident" (Powers); Knox,
(Racklift); Oxford, "breeds rarely" (Nash); Penobscot, "usually very
rare but often sporadically common, has been taken in late April"
(Knight); Piscataquis, "winter visitor, some winters common" (Homer);
Sagadahoc, "not common" (Spinney); Washington, uncertain, some
winters common, breeds in winter" (Boardman).

Genus ACANTHIS Bechstein.

219. (527a). Acanthis hornemannii exilipes (*Coues*). Hoary
Redpoll.

Dr. Brewer referred this species to eastern Maine (Cf. Brewer,
Proc. Boston Soc. Nat. Hist. Vol. 17, 1875, p. 441), but no evidence is there adduced to prove its occurrence. Verrill also mentions it in his list, but without citation of the grounds upon which
he admits it to the list. The first authentic record of this bird
having been taken in the state is given in the Proceedings of the
Portland Society of Natural History for April 1, 1897, by Mr. A.
H. Norton who writes: "A specimen of this rare bird was taken
at Westbrook, Maine, January 26, 1896. It was in a flock composed chiefly of common Redpolls and Pine Siskins, with a few
Greater Redpolls intermingled. It is a female, apparently not fully
mature." While it is quite probable that the species in question is

NOTE—Amadina rubronigra, an African species of Finch, has once been taken
in Maine though it was beyond doubt an escaped cage bird. (Cf. Allen, Bull.
Nutt. Orn. Club, Vol. 5, p. 120). Not knowing its proper place in the classification
I give it here.

of regular occurrence in winter, we have at present no evidence that such is the case.

220. (528). Acanthis linaria (*Linn.*). Redpoll.

A winter resident of irregular abundance throughout the state, and also a rare summer resident near Calais. Flocks of from 10 to 300 individuals may often be seen feeding on the seeds of alders, junipers, or various weeds by the roadsides and in the fields. Probably all four of the varieties sometimes occur in one flock.

County Records.—Androscoggin, "common winter visitor" (Johnson); Cumberland, "common winter migrant" (Mead); Franklin, "irregularly abundant winter visitor" (Lee and McLain); Hancock, "winter resident" (Dorr); Kennebec, "quite common winter visitor" (Powers); Knox, "winter" (Rackliff); Oxford, "visitant" (Nash); Penobscot, "irregularly abundant winter resident" (Knight); Piscataquis, "common winter visitor" (Homer); Sagadahoc, "common spring and fall" (Spinney); Somerset, "common winter resident" (Morrell); Waldo, (Spratt); Washington, "common winters, also summer resident" (Boardman); York, "migrant" (Adams).

221. (528 a). Acanthis linaria holbœllii (*Brehm*). Holbœll's Redpoll.

The first Maine and fifth eastern example of this bird is a male which was taken at North Bridgton, Cumberland County, on November 25, 1878, by Mr. J. C. Mead, who reports that it was in company with a flock of *A. linaria*. Mr. Mead sent the specimen to me along with a number of *A. linaria* for identification, and upon submitting it to Messrs. Brewster and Ridgway they agreed in referring it to this race. Mr. Mead has since very generously presented the specimen to me, and it now occupies a prominent place in my cabinet. The only other eastern examples recorded are one from Quebec (Ridgway), and three from Massachusetts (Brewster), but it may occur far more regularly than these records would indicate.

222. (528b). Acanthis linaria rostrata (Coues). Greater Redpoll.

The first published record of the occurrence of this species in the state is given by Prof. Wm. L. Powers in the Maine Sportsman for February, 1897, p. 9. This relates to a specimen shot by him at Gardiner, Kennebec County, on December 30, 1896. Mr. Fred Rackliff of Spruce Head has in his possession a specimen taken in

Knox County. It remained for Mr. A. H. Norton to show that this subspecies occurred abundantly near Westbrook in 1895. In the 6/ proceedings of the Portland Society of Natural History for April 1897, p. 104, he writes: "This large dark form was abundant in Westbrook during the months of January and February, 1895. 6/ It was first observed January 26, when it was less numerous than true *linaria* with which it was constantly associated. Its numbers were augmented by new arrivals, and on February 2nd it was the prevailing form. On the 8th of the month no Redpolls could be found. A return movement was soon noticed with constant increase in numbers until March 15th. *Rostrata* was not observed after February 27th." From this we may safely say that this race is likely to occur commonly in winter.

Genus SPINUS Koch.

223.   (529).   Spinus tristis (*Linn.*).   American Goldfinch.

A common summer resident throughout the state and also of not infrequent occurrence in winter. This species is commonly known as Yellow Bird, Wild Canary, and Thistle Bird.

County Records.—Androscoggin, "common summer resident" (Johnson); Aroostook, "common at Fort Fairfield" (Batchelder, Bull. Nutt. Orn. Club, Vol. 7, p. 147); Cumberland, "common resident" (Mead); Franklin, "common summer resident, sometimes in winter" (Lee and McLain); Hancock, "summer resident" (Murch); Kennebec, "common resident" (Gardiner Branch); Knox, "summer" (Rackliff); Oxford, "common, breeds" (Nash); Penobscot, "common in summer and rare winters" (Knight); Piscataquis, "common, often resident" (Homer); Sagadahoc, "common summer resident" (Spinney); Somerset, "common resident" (Morrell); Waldo, (Spratt); Washington, "abundant, breeds" (Boardman); York, "breeds" (Adams).

224.   (533).   Spinus pinus (*Wils.*).   Pine Siskin.

A common migrant in the fall, often sporadically abundant, and somewhat rarer in winter. It is a resident species in those parts of the state which are within the Canadian fauna.

County Records.—Androscoggin, "rare winter visitor" (Call); Aroostook, "seen at Sherman in June" (Knight); Cumberland, "common migrant, one nest taken" (Mead); Franklin, "common resident" (Swain); Hancock, "migrant" (Knight); Kennebec, "very rare" (Dill); Knox, "winter visitant" (Rackliff); Oxford, "breeds" (Nash); Penobscot, "common in fall, rare in winter and summer" (Knight); Piscataquis, "rare summer resident" (Whitman), "common winter resident"

(Homer); Somerset, "very irregular, sometimes summer resident" (Morrell); Waldo, "rare" (Spratt); Washington, "winter visitant, sometimes summer resident" (Boardman).

Genus PLECTROPHENAX Stejneger.

225. (534). Plectrophenax nivalis (*Linn.*). Snowflake.

Winter resident, everywhere abundant. These birds may be found in flocks running along the country roads, seeking for the undigested seeds in horse droppings. They also feed on various seeds in the fields and meadows, and are especially likely to be found about manure heaps in the rear of barns, when the snow has covered all other sources of food.

County Records.—Androscoggin, "common winter visitor" (Johnson); Cumberland, "common winter visitor" (Mead); Franklin, "winter visitor" (Swain); Hancock, "winter resident" (Dorr); Kennebec, "common winter resident" (Gardiner Brauch); Knox, "winter" (Rackliff); Oxford, "visitant" (Nash); Penobscot, "common November to April and often very abundant" (Knight); Piscataquis, "common winter visitor" (Homer); Sagadahoc, "common fall, spring and winter" (Spinney); Somerset, "common winter resident" (Morrell); Waldo, (Spratt); Washington, "winter visitant" (Boardman); York, "common" (Adams).

Genus CALCARIUS Bechstein.

226. (536). Calcarius lapponicus (*Linn.*). Lapland Longspur.

An irregular and rare winter visitor from the north.

County Records.—Cumberland, "very rare winter resident" (Brown's Cat. Birds of Portland, p. 13); Knox, "I have a specimen taken at St. George" (Norton); Oxford, "visitant" (Nash); Piscataquis, "rare" (Homer); Washington, "very rare" (Boardman).

227. (538). Calcarius ornatus (Towns.). Chestnut-collared Longspur.

A straggler from the west and of purely accidental occurrence. A specimen was taken at Scarborough. Cumberland County, August 13, 1886. (Cf. Goodale, The Auk, Vol. 4, p. 77).

Genus POOCÆTES Baird.

228. (540). Poocætes gramineus (*Gmel.*). Vesper Sparrow.

This bird is commonly known to rural observers as Grass Finch. It is common as a summer resident of fields and grassy meadows throughout the state.

County Records.—Androscoggin, "common summer resident" (Johnson); Aroostook, "common at Fort Fairfield" (Batchelder, Bull. Nutt. Orn. Club, Vol. 7, p. 148); Cumberland, "common summer resident" (Mead); Franklin, "common summer resident" (Swain); Hancock, "summer resident" (Murch); Kennebec, "abundant summer resident" (Gardiner Branch); Knox, "summer" (Racklift); Oxford, "common summer resident" (Johnson); Penobscot, "common summer resident" (Knight); Piscataquis, "common, breeds" (Homer); Sagadahoc, "common summer resident" (Spinney); Somerset, "quite common summer resident" (Morrell); Waldo, (Spratt); Washington, "abundant summer resident" (Boardman); York, "abundant summer resident" (Adams).

Genus AMMODRAMUS Swainson.

Subgenus PASSERCULUS Bonaparte.

229. (541). Ammodramus princeps (*Mayn.*). Ipswich Sparrow.

A somewhat rare migrant along the coast. This bird has at present only been known to breed on Sable Island, Nova Scotia, while elsewhere it occurs as a migrant or in winter. Its habits, nest, eggs, etc., are very minutely described by Dr. Dwight. (Cf. Dwight, The Ipswich Sparrow and Its Summer Home, Memoirs of the Nutt. Orn. Club, No. 2).

County Records.—Cumberland, "transient, rare in spring, common in autumn, confined to the seashore" (Brown's Cat. Birds of Portland, p· 13); Knox, "rare migrant" (Rackliff); Sagadahoc, "one specimen in spring" (Spinney).

230. (542 a). Ammodramus sandwichensis savanna (*Wils.*). Savanna Sparrow.

A common summer resident of most parts of the state. I have found it especially abundant on many of the grassy islands along the coast.

County Records.—Androscoggin, "fairly common summer resident" (Johnson); Aroostook, "common at Fort Fairfield" (Batchelder, Bull. Nutt. Orn. Club, Vol. 7, p. 148); Cumberland, "rare near Bridgton" (Mead), "abundant summer resident" (Brown's Cat. Birds of Portland, p. 13); Franklin, "common summer resident" (Richards); Hancock, "common summer resident especially on the islands" (Knight); Kennebec, "abundant summer resident" (Gardiner Branch); Knox, "summer" (Rackliff); Lincoln, "common on the islands" (Norton); Oxford, "fairly common summer resident" (Johnson); Penobscot, "common summer resident" (Knight); Piscataquis, "common, breeds" (Homer); Sagadahoc, "common summer resident" (Spinney);

Somerset, "common summer resident" (Morrell); Waldo, "common summer resident" (Knight); Washington, "abundant summer resident" (Boardman).

Subgenus COTURNICULUS Bonaparte.

231. (546). Ammodramus savannarum passerinus (*Wils.*). Grasshopper Sparrow.

An accidental visitor from the south of which there is only one record for the state. Mr. Boardman reports it as very rare or accidental at Calais, Washington County. (Cf. Boardman, Proc. Bost. Soc. Nat. Hist., Vol. 9, p. 126).

Subgenus AMMODRAMUS.

232. (549.) Ammodramus caudacutus (*Gmel.*). Sharp-tailed Sparrow.

A summer resident along the southern coast. Nathan Clifford Brown found this species at Scarboro late in October, 1876. (Cf. Brown, Bull. Nutt. Orn. Club, Vol. 2. p. 27; Vol. 3, p. 98). During 1879 he found them there in the summer and apparently breeding. (Cf. Brown, Bull. Nut. Orn. Club, Vol. 5, p. 52). In a recent paper on "The Sharp-tailed Finches of Maine," A. H. Norton says: "Though search has now been made, it has not been found farther to the north than Scarboro, Maine, and the physical features of the coast are such as to suggest the improbability of the normal range extending beyond this town." (Cf. Norton, Proc. Port. Soc. Nat. Hist., Vol. 2, March 15, 1897, p. 99). It has been recorded from Cumberland County only.

233. (549a). Ammodramus caudacutus nelsoni *Allen*. Nelson's Sparrow. Admitted on Mr. Norton's excellent authority as follows: "This bird, an inhabitant during the breeding season of the valleys of the Great Lakes, Upper Mississippi, Red River of the North, and portions of Missouri, is now added to the fauna of Maine on the strength of two specimens taken at Scarboro, October 16th, 1894. It is likely to be found in numbers all along the Maine coast." (Cf. Norton, Proc. Port. Soc. Nat. Hist., 1897, p. 99). He also says that specimens have been taken within the habitat of *caudacutus* near the breeding season, but there is no proof they were breeding.

7

234. (549b). Ammodramus caudacutus subvirgatus *Dwight*. Acadian Sharp-tailed Sparrow.

First taken in the state at Scarboro, Cumberland County in October, 1879. (Cf. Dwight, The Auk, Vol. 4, p. 237). I will here again quote from Mr. Norton's paper, cited previously, which is authoritative on the status of these Finches in the state. He says: "It therefor gives me much pleasure to have the privilege of introducing this bird as a summer resident on our coast, breeding in fair numbers as far to the southwestward as Small Point, Sagadahoc County. Here I observed them August 7, 1896, at which date they were still engaged in their domestic duties. The males were in full song, and in particular spots were to be heard quite constantly, a fact that I found to be of much importance in finding them, even while in their very midst." (Cf. Norton, ibid, p. 100). I regret that lack of space forbids me citing more of this extremely interesting paper. Mr. Boardman writes me that he has taken a Sharp-tailed Sparrow near Calais, Washington County. It is probably referable to this subspecies which undoubtedly occurs along our entire eastern coast as a summer resident.

235. (550). Ammodramus maritimus (*Wils.*). Seaside Sparrow.

Of purely accidental occurrence as a straggler from the south, a specimen having been taken at Shark Island. (Cf. Smith, Forest and Stream, December 18, 1884, p. 405).

Genus ZONOTRICHIA Swainson.

236. (554). Zonotrichia leucophrys (*Forst.*). White-crowned Sparrow.

A fairly common migrant of quite general distribution.

County Records.—Androscoggin, "fairly common migrant" (Johnson); Cumberland, "common migrant" (Mead); Franklin, "rare migrant" (Richards); Hancock, "rare" (Murch); Kennebec, "rare migrant" (Larrabee); Knox (Rackliff); Oxford, "visitant" (Nash); Penobscot, "fairly rare migrant" (Knight); Piscataquis, "common migrant" (Homer); Somerset, "rare migrant" (Morrell); Washington, "very rare" (Boardman); York, "migrant" (Adams).

237. (558). Zonotrichia albicollis (*Gmel.*). White-throated Sparrow.

A common summer resident of those counties within the Canadian fauna, while in the migrations it is abundant everywhere. It is commonly known as Peabody Bird from its well known spring call which sounds as if it were whistling "come, come, sow your pea, sow your pea, sow your pea."

County Records.—Androscoggin, "fairly common summer resident" (Johnson); Aroostook, "abundant at Fort Fairfield, nests" (Batchelder, Bull. Nutt. Orn. Club, Vol. 7, p. 148); Cumberland, "common summer resident" (Mead); Franklin, "common summer resident" (Swain); Hancock, "summer resident, common on the wooded islands"(Knight); Kennebec, "quite common summer resident" (Gardiner Branch); Knox, "summer" (Rackliff); Oxford, "common breeder" (Nash); Penobscot, "abundant migrant, common summer resident" (Knight); Piscataquis, "common, breeds" (Homer); Sagadahoc, "common migrant" (Spinney); Somerset, "common summer resident" (Morrell); Waldo, "common summer resident" (Knight); Washington, "abundant summer resident" (Boardman); York, "migrant" (Adams).

Genus SPIZELLA Bonaparte.

238. (559). Spizella monticola (*Gmel.*). Tree Sparrow.

A common migrant in fall and spring, while it is not rare to find specimens of this bird wintering in suitable localities throughout the state.

County Records.—Androscoggin, "common migrant" (Johnson); Cumberland, "common migrant" (Mead), "rather common winter resident" (Brown's Cat. Birds of Portland, p. 14); Franklin, "common winter resident" (Richards); Hancock, "migrant" (Murch); Kennebec, "quite common" (Gardiner Branch); Knox, "winter" (Rackliff); Oxford, (Nash); Penobscot, "common migrant, have taken specimens in January" (Knight); Piscataquis, "winter visitor" (Homer); Sagadahoc, (Spratt); Somerset, "common winter resident, most abundant in fall and spring" (Morrell); Waldo, (Spratt); Washington, "only in migrations" (Boardman); York, "migrant" (Adams).

239. (560). Spizella socialis (*Wils.*). Chipping Sparrow.

A common and in some localities abundant summer resident. It was rightly named *socialis*, as it seems to prefer to frequent the neighborhood of dwellings when it is possible to do so, although I have found the species nesting quite a distance from any house. It is commonly known as Chippy and Hair Bird, this latter name being due to the fact that it almost invariably lines its nest with hair.

County Records.—Androscoggin, "abundant summer resident" (Johnson); Aroostook, "rather common at Fort Fairfield" (Batchelder, Bull. Nutt. Oru. Club, Vol. 7, p. 148); Cumberland, "common summer resident" (Mead); Franklin, "common summer resident" (Swain); Hancock, "summer resident" (Murch); Kennebec, "quite common summer resident" (Gardiner Branch); Knox, "summer" (Rackliff); Oxford, "common breeder" (Nash); Penobscot, "abundant summer resident" (Knight); Piscataquis, "common, breeds" (Homer); Sagadahoc, "common summer resident" (Spinney); Somerset, "common summer resident" (Morrell); Waldo, (Spratt); Washington, "very abundant summer resident" (Boardman); York, "abundant" (Adams).

240. (563). Spizella pusilla (*Wils.*). Field Sparrow.

Occurs as an uncommon summer resident of the Alleghanian fauna, while in the counties of the Canadian it is very rare or accidental.

County Records.—Androscoggin, "fairly common summer resident" (Call); Cumberland, "uncommon summer resident" (Brown's Cat. Birds of Portland, p. 14), "not common" (Mead); Franklin, "rare summer resident" (Swain); Kennebec, "summer resident" (Larrabee); Knox, "summer" (Norton); Oxford, "breeds" (Nash); Sagadahoc, "rare" (Spratt); York, "not common summer resident" (Adams.)

Genus JUNCO Wagler.

241. (567). Junco hyemalis (*Linn.*). Slate-colored Junco.

A common summer resident within the Canadian fauna, while elsewhere it is of common occurrence as a migrant, and also to a limited extent as a winter resident. Known to many under the name of Black Snowbird.

County Records.—Androscoggin, "abundant migrant, rare summer resident" (Johnson); Aroostook, "common at Fort Fairfield" (Batchelder, Bull. Nutt. Oru. Club, Vol. 7, p. 148); Cumberland, "common summer resident" (Mead), "abundant transient, uncommon summer resident, occasionally found throughout the winter" (Brown's Cat. Birds of Portland, p. 14); Franklin, "common resident" (Swain); Hancock, "summer resident" (Knight); Kennebec, "abundant" (Gardiner Branch); Knox, "resident" (Rackliff); Oxford, "common breeder" (Nash); Penobscot, "abundant migrant, not uncommon resident" (Knight); Piscataquis, "common, breeds" (Homer); Sagadahoc, (Spratt); Somerset, "common migrant, possibly summer resident" (Morrell); Waldo, "summer resident" (Spratt); Washington, "very abundant summer resident" (Boardman); York, "common migrant, a nest found in '82" (Adams).

Genus MELOSPIZA Baird.

242. (581). Melospiza fasciata (*Gmel.*). Song Sparrow.

An abundant summer resident everywhere, both on the islands of the coast and throughout the interior. Specimens have been taken in winter, but it is doubtful if the species is a regular winter resident, even in the extreme southern counties.

County Records.—Androscoggin, "abundant summer resident" (Johnson); Aroostook, "common at Sherman" (Knight); Cumberland, "common summer resident" (Mead); Franklin, "common summer resident" (Swain); Hancock, "summer resident, common on the islands" (Knight); Kennebec, "quite common summer resident" (Gardiner Branch); Knox, "summer resident" (Racklift); Oxford, "breeds commonly" (Nash); Penobscot, "common in summer, have seen it in February" (Knight); Piscataquis, "common, breeds" (Homer); Sagadahoc, "common summer resident" (Spinney); Somerset, "common summer resident" (Morrell); Waldo, "summer resident" (Spratt); Washington, "very abundant summer resident" (Boardman); York, "common summer resident" (Adams).

243. (583). Melospiza lincolnii (*Aud.*). Lincoln's Sparrow.

A rare spring and fall migrant which probably occurs throughout the state. Owing to its resemblance to the Song Sparrow, it is liable to escape detection by being mistaken for this species. A female specimen was shot at Westbrook, Cumberland County, September 20, 1896, by Arthur H. Norton. Mr. Boardman gives it as rare for Washington County and occurring in spring only.

244. (584). Melospiza georgiana (*Lath.*). Swamp Sparrow.

A fairly common summer resident of quite general distribution, but very likely to escape observation on account of its general resemblance to the other Sparrows.

County Records.—Androscoggin, "fairly common summer resident" (Johnson); Aroostook, "not common at Houlton" (Batchelder, Bull. Nutt. Orn. Club, Vol. 7, p. 148); Cumberland, "common summer resident" (Mead); Franklin, "quite common summer resident" (Lee and McLain); Kennebec, "rare" (Gardiner Branch); Knox, "summer" (Racklift); Oxford, "rare summer resident" (Johnson); Penobscot, "summer resident, commoner than one would believe from the very few specimens taken" (Knight); Somerset, "common summer resident" (Morrell); Washington, "not uncommon summer resident" (Boardman); York, "not common migrant" (Adams).

Genus PASSERELLA Swainson.

245. (585). Passerella iliaca (*Merr.*). Fox Sparrow.

Of quite general and common occurrence as a fall and spring migrant. This is the handsomest of our Sparrows and bears a not distant superficial resemblance to the Thrushes. At first glance I have often mistaken one of these birds for a Thrush, and it needed the second look to convince me of my mistake. They greatly love to scratch about among dead leaves and other fallen rubbish, while the noise thus produced is worthy of a larger bird. While hunting in the fall I have often mistaken the scratching of this bird among the leaves for a Grouse running along, and my gun would leap to my shoulder before the mistake in the identity of the noise producer was discovered.

County Records.—Androscoggin, "common migrant" (Johnson); Cumberland, "common migrant" (Mead); Franklin, "rare migrant" (Swain); Kennebec, "very rare migrant" (Powers); Knox, "migrant" (Rackliff); Oxford, "migrant, very common" (Nash); Penobscot, "common migrant" (Knight); Piscataquis, "common migrant" (Homer); Sagadahoc, "common spring migrant" (Spinney); Somerset, "common migrant" (Morrell); Washington, "fall and spring" (Boardman); York, "not very common migrant" (Adams).

Genus PIPILO Vieillot.

246. (587). Pipilo erythrophthalmus (*Linn.*). Towhee.

A somewhat common summer resident of those countries within Alleghanian fauna, while elsewhere in the state it is of rare or casual occurrence.

County Records.—Androscoggin, "rare summer resident" (Johnson); Cumberland, "rare" (Mead); Oxford, "breeds commonly" (Nash); Sagadahoc, "common summer resident" (Spinney); York, "very abundant" (Adams).

Genus CARDINALIS Bonaparte.

247. (593). Cardinalis cardinalis (*Linn.*). Cardinal.

It is highly probable that all of this species which have been taken in the state are escaped cage birds, but at the same time the fact that Cardinals have been taken in a state of freedom entitles them to be represented in the list. Prof. Powers of Gardiner writes that one of these birds was shot from a flock of three, in that vicinity, in 1896. In response to further queries on my part, he positively states that the specimen showed no signs of having been a caged bird. Smith's List also gives this species a place as a straggler or escaped cage bird. Whatever the manner of its occurrence, it must rank as an accidental visitor to the state, being purely a bird of the Carolinian fauna.

Genus ZAMELODIA Coues.

248. (595). Zamelodia ludoviciana (*Linn.*). Rose-breasted Grosbeak.

A rare summer resident of the eastern parts of the state, while elsewhere it is common, or even quite abundant in some localities.

County Records.—Androscoggin, "fairly common summer resident" (Johnson); Aroostook, "rather common at Fort Fairfield" (Batchelder, Bull. Nutt. Orn. Club, Vol. 7, p. 148); Cumberland, "common summer resident" (Mead); Franklin, "common summer resident" (Richards); Hancock, "rare" (Dorr); Kennebec, "common summer resident" (Gardiner Branch); Knox, "rare migrant" (Rackliff); Oxford, "common breeder" (Nash); Penobscot, "regular but rare summer resident" (Knight); Piscataquis, "not uncommon, breeds" (Homer); Sagadahoc, "five specimens" (Spinney); Somerset, "not common summer resident" (Morrell); Waldo, (Spratt); Washington, "rare summer resident" (Boardman); York, "quite abundant" (Adams).

Genus PASSERINA Vieillot.

249. (598). Passerina cyanea (*Linn.*). Indigo Bunting.

A not uncommon summer resident of quite general distribution within our limits.

County Records.—Androscoggin, "fairly common summer resident" (Johnson); Cumberland, "common summer resident" (Mead); Franklin, "common summer resident" (Swain); Hancock, "summer resident" (Dorr); Kennebec, "common summer resident" (Gardiner Branch); Knox, "rare migrant" (Rackliff); Oxford, "fairly common summer resident" (Johnson); Penobscot, "not very common summer resident" (Knight); Piscataquis, "summer resident, not common" (Homer); Sagadahoc, "not common, seen in June" (Spinney); Somerset, "rare summer resident" (Morrell); Washington, "not uncommon summer resident" (Boardman); York, "quite abundant, breeds" (Adams).

Genus SPIZA Bonaparte.

250. (604). Spiza americana (*Gmel.*). Dickcissel.

Of purely accidental occurrence as a straggler from the south. A specimen taken October 10, 1888, is recorded from Westbrook,

Cumberland County, by Mr. Norton. (Cf. Norton, The Auk, Vol. 10, p. 302, and also ibid. Vol. 11, pp. 78-79). Mr. Rackliff reports taking a specimen in Knox County. A third example has been taken on Job's Island, Penobscot Bay. (Cf. Townsend, The Auk, Vol. 2, p. 106).

Family TANAGRIDÆ. Tanagers.

Genus PIRANGA Vieillot.

251. (607). Piranga ludoviciana (*Wils.*). Louisiana Tanager.

Accidental, a specimen having been taken near Bangor about October 1, 1889, and sent to Mr. S. L. Crosby, the well known Bangor taxidermist. Regarding this same bird Mr. Manly Hardy writes: "I saw the remains of that Louisiana Tanager you ask about. It was an adult male and was brought in the flesh but too far gone to mount. It was a Louisiana Tanager without any question as I compared it with mine." The fact that this was compared with named specimens by such a reliable and careful observer as Mr. Hardy is a sufficient voucher for the reliability of this record.

252. (608). Piranga erythromelas *Vieill.* Scarlet Tanager.

A rare summer resident of quite general distribution within our limits. The handsome scarlet males, with black wings and tail, are easily identified by any persons having the slightest knowledge of Ornithology, but the duller colored females, while easily identified, have not the prominent colors of their mates.

County Records.—Androscoggin, "rare migrant" (Johnson); Aroostook, "rare at Houlton" (Batchelder, Bull. Nutt. Orn. Club, Vol. 7, p. 111); Cumberland, "rare" (Mead); Franklin, "rare summer resident" (Swain); Hancock, "summer resident" (Dorr); Kennebec, "very rare summer resident" (Robbins); Knox, "rare migrant" (Rackliff); Oxford, "breeds" (Nash); Penobscot, "rare, not seen or reported since 1891" (Knight); Piscataquis, "rare, breeds" (Homer); Sagadahoc, "four specimens in spring" (Spinney); Waldo, "rare" (Spratt); Washington, "rare summer resident" (Boardman); York, "rare, sometimes breeds" (Adams).

253. (610). Piranga rubra (*Linn.*). Summer Tanager.

There is seemingly but one record for the state, this specimen being taken at Wiscasset, Lincoln County, and recorded in Smith's List in the Forest and Stream. Mr. Boardman has taken it in New Brunswick.

Family HIRUNDINIDÆ. Swallows.

Genus PROGNE Boie.

254. (611). Progne subis (*Linn.*). Purple Martin.

A common summer resident in the vicinity of dwellings where martin houses have been erected for their accommodation. They seemingly return to the same house every year, and usually have a hard fight to regain possession of it, as during their absence the English Sparrows have usually taken possession. However, in such fights the Martins usually prove victorious, and the intruding *Passer domesticus* is forced to seek a new home.

County Records.—Androscoggin, "abundant summer resident" (Johnson); Aroostook, "seen at Fort Fairfield and Houlton" (Batchelder, Bull. Nutt. Orn. Club, Vol. 7, p. 110); Cumberland, "common summer resident" (Mead); Franklin, "common summer resident" (Swain); Hancock, "common throughout the summer" (Knight); Kennebec, "very rare summer resident" (Larrabee); Knox, "summer" (Rackliff); Oxford, "common breeder" (Nash); Penobscot, "common in the cities, somewhat rarer in the country, but occur wherever houses have been provided for their benefit" (Knight); Piscataquis, "common, breeds" (Homer); Sagadahoc (Spratt); Somerset, "common summer resident" (Morrell); Waldo (Spratt); Washington, "common summer resident" (Boardman); York, "common summer resident" (Adams).

Genus PETROCHELIDON Cabanis.

255. (612). Petrochelidon lunifrons (*Say*). Cliff Swallow.

A very common summer resident of general distribution. These birds are commonly called Eave Swallows from their habit of building their large flask-shaped nests of mud beneath the eaves of buildings. They are also called Republicans, presumably because they nest in colonies.

County Records.—Androscoggin, "common summer resident" (Johnson); Aroostook, "abundant at Fort Fairfield" (Batchelder, Bull. Nutt. Orn. Club, Vol. 7, p. 110); Cumberland, "common summer resident" (Mead); Franklin "common summer resident" (Swain); Hancock, "common on the inhabited islands along the coast and also in the interior, breeds, (Knight); Kennebec, "very common summer resident" (Gardiner Branch); Knox, "summer" (Rackliff); Oxford, "common, breeds" (Nash); Penobscot, "abundant summer resident" (Knight); Piscataquis, "common, breeds" (Homer); Sagadahoc, "common summer resident" (Spinney); Somerset, "common summer resident" (Morrell); Waldo (Spratt); Washington, "very abundant summer resident" (Boardman); York, "common summer resident" (Adams).

Genus CHELIDON Forster.

256. (613).   Chelidon erythrogastra (*Bodd.*).   Barn Swallow.

A common summer resident throughout the state. These birds nearly always nest in colonies, placing their nests of mud within barns and unoccupied houses, and attaching them to the side or placing them on top of some beam. They are very common on some of the islands along our coast, and in July, 1893, I found two pair nesting in an unoccupied hut on Seal Island which is situated far out to sea.

County Records —Androscoggin, "abundant summer resident" (Johnson); Aroostook, "common at Fort Fairfield" (Batchelder, Bull. Nutt. Orn. Club, Vol. 7, p. 110); Cumberland, "common summer resident" (Mead); Franklin, "common summer resident" (Swain); Hancock, "breeds on many of the inhabited islands and in the interior" (Knight); Kennebec, "quite common summer resident" (Gardiner Branch); Knox, "summer" (Racklift); Oxford, "breeds commonly" (Nash); Penobscot, "abundant summer resident" (Knight); Piscataquis, "common, breeds" (Homer); Sagadahoc, "common summer resident" (Spinney); Somerset, "common summer resident" (Morrell); Waldo, (Spratt); Washington, "very abundant summer resident" (Boardman); York, "common summer resident" (Adams).

Genus TACHYCINETA Cabanis.

257. (614).   Tachycineta bicolor (*Vieill.*).   Tree Swallow.

A common summer resident both in the vicinity of houses and in the wilderness. Near civilization it prefers to place its nest in some hole or crevice of a building or in an unoccupied martin house, while in other localities it nests in holes in trees. I have found these birds especially common along our rivers and about our ponds and lakes, in such places placing their nests in deserted woodpecker or other holes, in stumps near to the water.

County Records.—Androscoggin, "tolerably common summer resident" (Johnson); Aroostook, "abundant at Fort Fairfield" (Batchelder, Bull. Nutt. Orn. Club, Vol. 7, p. 110); Cumberland, "common summer resident" (Mead); Franklin, "common summer resident" (Swain); Hancock, "common summer resident" (Knight); Kennebec, "common summer resident" (Gardiner Branch); Knox, "summer" (Racklift); Oxford, "common, breeds" (Nash); Penobscot, "breeds commonly" (Knight); Piscataquis, "common, breeds" (Homer); Sagadahoc, "common summer resident" (Spinney); Somerset, "common summer resident" (Morrell); Waldo, (Spratt); Washington, "very abundant summer resident" (Boardman); York, "common summer resident" (Adams).

Genus CLIVICOLA Forster.

258. (616). Clivicola riparia (*Linn.*). Bank Swallow.

Common summer resident in localities where the sand banks afford perpendicular walls in which these birds can excavate their nesting burrows. These are often dug to a depth of three feet, although the average in places where the birds are not disturbed by small boys is about a foot and a half. At the end of these, the four to seven white eggs are deposited in a rudely made nest of dry grass or straw, which is often lined with feathers.

County Records.—Androscoggin, "abundant summer resident" (Johnson); Aroostook, "common at Fort Fairfield" (Batchelder, Bull. Nutt. Orn. Club, Vol. 7, p. 110); Cumberland, "common summer resident" (Mead); Franklin, "common summer resident" (Swain); Hancock, "I have found this species nesting abundantly along the shores of many islands along the coast" (Knight); Kennebec, "common summer resident" (Gardiner Branch); Knox, "summer" (Racklift); Lincoln, "breeds at Damariscotta" (H. E. Berry, The Oologist, December, 1888, p. 175); Oxford, "common breeder" (Nash); Penobscot, "abundant breeder" (Knight ; Piscataquis, "common, breeds" (Homer); Sagadahoc, "common summer resident" (Spinney); Somerset, "common summer resident (Morrell); Waldo, "common summer resident" (Knight); Washington, "very abundant summer resident" (Boardman); York, "common summer resident" (Adams).

Family AMPELIDÆ. Waxwings etc.

Subfamily AMPELINÆ. Waxwings.

Genus AMPELIS Linnæus.

259. (618). Ampelis garrulus *Linn.* Bohemian Waxwing.

An irregular winter visitor from the north which probably occurs throughout the entire state, although it has only been recorded from two counties.

County Records.—Kennebec, given in Hamlin's list of "Birds of Waterville," Report of Secretary Maine Board of Agriculture, 1865, pp. 168-173) ; Washington, "rare. some winters occurs in large flocks" (Boardman).

260. (619.). Ampelis cedrorum (*Vieill.*). Cedar Waxwing.

A common summer resident and of rare occurrence in winter. Commonly known as Cherry Bird on account of their fondness for this fruit. Upon their arrival in the spring I have often observed them engaged in pecking at apple blossoms, and seemingly eating parts of the same, though I have never shot one at this period, so engaged, and cannot say what part of the blossom, if any, is actually eaten. At this time of the year the farmers call them Apple Birds and this term applies until the advent of ripe cherries brings a change in their diet and name.   They are certainly insectivorous to a large extent, and undoubtedly devour enough injurious insects to more than pay for the limited quantities of fruit they take as toll.

County Records.—Androscoggin, "common summer resident" (Johnson); Aroostook, "common" (Batchelder, Bull. Nutt. Orn. Club, Vol. 7, p. 110); Cumberland, "common summer resident" (Mead); Franklin, "common summer resident" (Swain); Hancock, "summer resident" (Murch); Kennebec, "common" (Gardiner Branch); Knox, "summer" (Racklift); Oxford, "common breeder" (Nash); Penobscot, "common summer resident, rarely seen in winter" (Knight); Piscataquis, "common, breeds" (Homer); Sagadahoc, "common summer resident" (Spinney); Somerset, "common summer resident" (Morrell); Waldo, (Spratt); Washington, "common summer resident, some in winter" (Boardman); York, "common summer resident" (Adams).

Family LANIIDÆ.   Shrikes.
Genus Lanius Linnæus.

261. (621).   Lanius borealis *Vieill.*   Northern Shrike.

Of quite common occurrence as a winter resident. This species does not breed in the state, all published records to the contrary notwithstanding.   All statements that this species has been found nesting in the state are made by incompetent observers, and upon investigation will be found to refer to the succeeding species.

County Records.—Androscoggin, "fairly common winter resident" (Johnson); Cumberland, "common winter migrant" (Mead); Franklin, "rare winter resident" (Swain); Hancock, (Dorr); Kennebec, "rare" (Gardiner Branch); Knox, "winter" (Racklift); Oxford, "fairly common migrant" (Johnson); Penobscot, "quite common in late fall, winter and early spring" (Knight); Piscataquis, (Homer); Sagadahoc, "common in winter" (Spinney); Somerset, "not common winter visitant" (Morrell); Washington, "common fall and winter" (Boardman); York, "rare migrant" (Adams).

262.   (622).   Lanius ludovicianus *Linn.*   Loggerhead Shrike.

Summer resident in many localities where the conditions are favorable, while in other places the species has not been reported. In common with the preceding it is called Butcher Bird, and I have heard the name Joree also applied to it, this latter coming from an attempt to syllabize the cry of the bird.   Our Maine birds are intermediate between the Loggerhead and White-rumped varieties, but on the whole they approach nearest to the former, and have been assigned to it whenever specimens have been sent to authorities for identification. The A. O. U. Check-List also gives New England as part of its habitat.

County Records.—Androscoggin, "rare summer resident" (Johnson); Cumberland, "of regular occurrence, it has come to my notice only during April and August, in Westbrook, Gorham, etc." (Norton); Franklin, "rare summer resident" (Richards); Hancock, "summer resident" (Murch); Kennebec, "rare" (Gardiner Branch); Oxford, "rare summer resident" (Johnson); Penobscot, "common summer resident" (Knight); Piscataquis, "common summer resident" (Whitman); Somerset, "quite common summer resident" (Morrell); Washington, "rare summer resident" (Boardman); York, "rare migrant" (Adams).

Family VIREONIDÆ. Vireos.

Genus VIREO Vieillot.

Subgenus VIREOSYLVA Bonaparte.

263. (624). Vireo olivaceus (*Linn.*). Red-eyed Vireo.

The commonest species of its family with us, and of very general distribution as a summer resident within our limits. It is a bird of both woodland and shady city streets, its presence during the breeding season being evidenced by its ever constant song.

County Records.—Androscoggin, "common summer resident" (Johnson); Aroostook, "common at Fort Fairfield" (Batchelder, Bull. Nutt. Orn. Club, Vol. 7, p. 111); Cumberland, "common summer resident" (Mead); Franklin, "common summer resident" (Swain); Hancock, "summer resident" (Murch); Kennebec, "common summer resident" (Gardiner Branch); Knox, "summer" (Rackliff); Oxford, "breeds commonly" (Nash); Penobscot, "common breeder" (Knight); Piscataquis, "common, breeds" (Homer); Sagadahoc, "common summer resident" (Spinney); Somerset, "common summer resident" (Morrell); Waldo, (Spratt); Washington, "very abundant summer resident" (Boardman); York, "common summer resident" (Adams).

264. (626). Vireo philadelphicus (*Cass.*). Philadelphia Vireo.

A summer resident of the Canadian fauna, while elsewhere in the state it is of somewhat rare occurrence as a migrant, according to the data now at hand. It is very probable that this species has been overlooked by many Ornithologists, owing to its resemblance to the Warbling species, while its song is almost indistinguishable from that of the Red-eyed Vireo.

County Records.—Franklin, "rare migrant" (Richards); Kennebec, "occurs at Waterville" (Deane, Bull. Nutt. Orn. Club, Vol. 1, p. 74); Oxford, "at the 1896 Congress of the A. O. U. Mr. Brewster spoke of this bird being observed at Upton in the breeding season and that it was fairly common"; Washington, "rare" (Boardman).

265. (627). Vireo gilvus (*Vieill.*). Warbling Vireo.

A fairly common summer resident of many favored localities within the state, and seemingly showing a marked partiality for the various shade trees which line the streets of our cities and towns. It is rightly named Warbling Vireo as its rolling warbling song may be heard wherever it occurs during the breeding season.

County Records.—Androscoggin, "fairly common summer resident" (Johnson); Cumberland, "rare" (Mead); Franklin, "rare summer resident" (Swain); Hancock, "summer resident" (Dorr); Kennebec, "common summer resident" (Gardiner Branch); Oxford, "occurs at Norway" (Purdie, Bull. Nutt. Orn. Club, Vol. 2, p. 15); Penobscot, "rare summer resident within City of Bangor along the shaded streets" (Knight); Somerset, "not common summer resident" (Morrell); Washington, "not plenty, summer resident" (Boardman).

Subgenus LANIVIREO Baird.

266. (628). Vireo flavifrons *Vieill.* Yellow-throated Vireo.

This is seemingly the rarest of our Vireos, with the possible exception of *V. philadelphicus*, and like the rest of its family it is a summer resident within our boundaries.

County Records.—Androscoggin, "rare summer resident" (Johnson); Cumberland, "rare, two specimens, taken in May and on July 31, 1878, near Bridgton" (Mead), "I know of but one specimen which was taken May 21, 1881" (Brown's Cat. Birds of Portland, p. 10); Franklin, "rare summer resident" (Swain); Kennebec, (Robbins); Sagadahoc, "rare" (Spratt).

267. (629). Vireo solitarius (*Wils.*). Blue-headed Vireo.

Of quite rare and somewhat local distribution during the breeding season, while in some places it is of fairly common occurrence as a migrant. It is to be looked for in the depths of the woods, hence the name Solitary Vireo which is often applied to it.

County Records.—Androscoggin, "rare summer resident" (Johnson); Aroostook, "common at Houlton, not common at Fort Fairfield" (Batchelder). (Bull. Nutt. Orn. Club, Vol. 7, p. 111); Cumberland, "common migrant" (Mead); "rather rare summer resident" (Brown's Cat. Birds of Portland, p. 10); Franklin, "rare summer resident" (Swain); Kennebec, "very rare" (Gardiner Branch); Knox, "summer resident" (Norton); Oxford, "occurs at Upton" (Brewster Bull. Nutt. Orn. Club, Vol. 3, p. 116); Penobscot, "rare summer resident" (Knight); Piscataquis, "rare" (Homer); Sagadahoc, (Spratt); Somerset, "not common migrant" (Morrell); Waldo, (Spratt); Washington, "not common summer resident" (Boardman.)

Family MNIOTILTIDÆ. Wood Warblers.
Genus MNIOTILTA Vieillot.

268. (636). Mniotilta varia (*Linn.*). Black and White Warbler.

Of quite general occurrence as a summer resident and common during the migrations. This bird much resembles the Creepers in habits, creeping up and down tree trunks in search of food, hanging head down or in other seemingly impossible positions, and acting entirely different from the other Warblers.

County Records.—Androscoggin, "fairly common summer resident" (Johnson); Aroostook, "observed at Fort Fairfield" (Batchelder, Bull. Nutt. Orn. Club, Vol. 7, p. 109); Cumberland, "common summer resident" (Mead); Franklin, "common summer resident" (Swain); Hancock, "quite common summer resident" (Knight); Kennebec, "quite common summer resident" (Gardiner Branch); Knox, "summer resident" (Norton); Oxford, "breeds commonly" (Nash); Penobscot, "common migrant and fairly common summer resident" (Knight); Piscataquis, "common summer resident" (Whitman); Sagadahoc, "common migrant" (Spinney); Somerset, "common summer resident" (Morrell); Waldo, (Spratt); Washington, "common summer resident" (Boardman); York, "very common summer resident" (Adams).

Genus PROTONOTARIA Baird.

269. (637). Protonotaria citrea (*Bodd.*). Prothonotary Warbler.

Accidental, a single individual having been taken at Calais, Washington County, October 30th, 1862, by Mr. Boardman. (Cf. Brewster, Bull. Nutt. Orn. Club, Vol. 3, p. 153).

Genus HELMINTHOPHILA Ridgway.

270. (645). Helminthophila rubricapilla (*Wils.*). Nashville Warbler.

A very common migrant and fairly common summer resident of most portions of the state.

County Records.—Androscoggin, "rare summer resident" (Johnson);
Aroostook, "observed at Fort Fairfield" (Batchelder, Bull. Nutt. Orn.
Club, Vol. 7, p. 110); Cumberland, "common migrant" (Mead), "common summer resident" (Brown's Cat. Birds of Portland, p. 6); Franklin,
"common summer resident" (Swain); Hancock, "summer resident"
(Knight); Kennebec, "common summer resident" (Larrabee); Knox,
"summer" (Racklift); Oxford, "breeds" (Nash); Penobscot, "quite common summer resident, abundant in migrations" (Knight); Piscataquis,
"not common, breeds" (Homer); Sagadahoc, (Spratt); Somerset, "quite
common summer resident" (Morrell); Washington, "common summer
resident" (Boardman).

271. (647). Helminthophila peregrina (*Wils.*). Tennessee
Warbler.

A quite rare summer resident within the Canadian fauna and of
occurrence elsewhere chiefly as a migrant. Owing to its close
resemblance to the Nashville Warbler, this bird has probably been
overlooked by many observers.

County Records.—Androscoggin, "rare summer resident" (Johnson);
Franklin, "rare migrant" (Richards); Kennebec, (Larrabee); Oxford,
"breeds at Upton" (Mayard's List of Birds of Coos Co., N. H., and
Oxford Co., Maine, p. 7); Penobscot, "very rare summer resident"
(Knight); Piscataquis, "rare summer resident" (Whitman); Somerset,
"one shot out of a flock of six or eight, May 15, 1896" (Morrell);
Washington, "common summer resident" (Boardman).

Genus COMPSOTHLYPIS Cabanis.

272. (648a). Compsothlypis americana usneæ *Brewster*.
Northern Parula Warbler.

This new described subspecies of the Parula or Blue-Yellow-
backed Warbler is a fairly common summer resident of many portions of the state. The nest is seemingly always placed in a clus-
ter of the usnea lichen, (*Usnea longissima*) and usually at no
great distance from the ground.

County Records.—Androscoggin, "rare summer resident" (Johnson);
Cumberland, "common summer resident" (Mead); Franklin, "rare sum-
mer resident" (Swain); Kennebec, "common summer resident" (Gar-
diner Branch); Knox, "summer resident" (Norton); Oxford, "Breeds at
Upton" (Mayard's List of Birds of Coos Co., N. H., and Oxford, Co.,
Maine, p. 6); Penobscot, "quite common summer resident" (Knight);
Piscataquis, "common" (Homer); Sagadahoc, "common summer resi-
dent" (Spinney); Somerset, "common migrant, apparently not a sum-
mer resident" (Morrell); Washington, "not uncommon summer resident"
(Boardman).

Genus DENDROICA Gray.
Subgenus PERISSOGLOSSA Baird.

273. (650). Dendroica tigrina (*Gmel.*). Cape May Warbler.

A somewhat uncommon summer resident of the Canadian fauna, elsewhere of quite rare occurrence in migrations.

County Records.—Androscoggin, "rare migrant" (Johnson); Aroostook, "a male shot at Fort Fairfield" (Batchelder, Bull. Nutt. Orn. Club, Vol. 7, p. 110); Cumberland, "common migrant" (Mead); Franklin, "rare migrant" (Richards); Kennebec, (Gardiner Branch); Oxford, "probably breeds" (Given in Mayard's List of Birds of Coos Co., N. H., and Oxford Co., Maine, p. 13); Piscataquis, "rare" (Homer); Sagadahoc, "rare" (Spratt); Somerset, "rare, one specimen taken August 22, 1893" (Morrell); Washington, "summer resident of variable abundance" (Boardman).

Subgenus DENDROICA Gray.

274. (652). Dendroica æstiva (*Gmel.*). Yellow Warbler.

The Summer Yellow-bird is one of our commonest Warblers, being a very common summer resident of general distribution. It is characteristic of no particular faunal area, being found throughout all temperate North America, except in the extreme southwest and northwest where its subspecies occur.

County Records.—Androscoggin, "common summer resident" (Johnson); Aroostook, "occurs at Fort Fairfield" (Batchelder, Bull. Nutt. Orn. Club, Vol. 7, p. 109); Cumberland, "common summer resident" (Mead); Franklin, "common summer resident" (Swain); Hancock, "summer resident" (Murch); Kennebec, "abundant summer resident" (Gardiner Branch); Knox, "summer" (Rackliff); Oxford, "abundant at Norway" (Verrill's List of the Birds of Norway, Proc. Essex Inst., Vol. 3, p. 136 et seq.); Penobscot, "very common summer resident" (Knight); Piscataquis, "rare summer resident" (Homer); Sagadahoc "common summer resident" (Spinney); Somerset, "summer resident" (Morrell); Waldo (Spratt); Washington, "abundant summer resident" (Boardman); York, "quite common summer resident" (Adams).

275. (654). Dendroica cærulescens (*Gmel.*). Black-throated Blue Warbler.

A somewhat common summer resident in some sections of the state, and of very general occurrence as a migrant. The males being of a general grayish-blue color above, the sides of their head and the throat black, and their breast and belly white, together

8

with the white spot on the wing at the end of the primary wing
coverts, they are easily identified. The females are more sober in
coloration, though equally easily recognized by persons acquainted
with the species.

County Records.—Androscoggin, "rare summer resident" (Johnson);
Aroostook, "rather common at Fort Fairfield" (Batchelder, Bull. Nutt.
Orn. Club, Vol. 7, p. 109); Cumberland, "common migrant" (Mead);
Franklin, "rare summer resident" (Swain); Kennebec, "summer resident" (Gardiner Branch); Knox, "rare migrant" (Rackliff); Oxford,
"common and breeding at Upton" (Maynard's List of Birds of Coos Co.,
N. H., and Oxford Co., Me., p. 8); Penobscot, "common migrant
and summer resident" (Knight); Piscataquis, "common" (Homer);
Sagadahoc, "three specimens, all in fall" (Spinney); Somerset, "rare
summer resident" (Morrell); Washington, "not abundant summer resident" (Boardman); York, "migrant" (Adams).

276. (655). Dendroica coronata (*Linn.*). Myrtle Warbler.

Of very general distribution and everywhere common in the
migrations, being known to many persons as the Yellow Rumped
Warbler. It is also a common summer resident in the northern
and eastern parts of the state, growing less abundant at this season
as one enters the counties of the Alleghanian fauna. It is primarily
a bird whose distribution in the breeding season is limited by the
southern boundaries of the Canadian fauna, although a few stragglers remain to nest south of this limit. It is the first Warbler to
appear in spring and the last to leave in fall.

County Records.—Androscoggin, "fairly common summer resident"
(Johnson); Aroostook, "common at Fort Fairfield" (Batchelder, Bull.
Nutt. Orn. Club, Vol. 7, p. 109); Cumberland, "abundant transient near
Portland, six individuals were seen at Pine Point on January 1, 1885,
and two of them secured" (Brown's Cat. Birds of Portland, pp. 7 and 38);
"common summer resident" (Mead); Franklin, "common summer resident" (Swain); Hancock, "common summer resident, occurs at this
season on many of the wooded islands along the coast" (Knight); Kennebec, "common summer resident" (Larrabee); Knox, "summer" (Rackliff); Oxford, "common at Upton in the breeding season" (Maynard's
List of Birds of Coos Co., N. H., and Oxford Co., Me., p. 8); Penobscot,
"common summer resident" (Knight); Piscataquis, "common, breeds"
(Homer); Sagadahoc, "common summer resident" (Spinney); Somerset,
"rare summer resident, common migrant" (Morrell); Waldo, "common
summer resident" (Knight); Washington, "abundant summer resident"
(Boardman); York, "quite common" (Adams).

277. (657). Dendroica maculosa (*Gmel.*). Magnolia Warbler. Known to many as the Black and Yellow Warbler, this species is a quite common summer resident of general distribution. While seemingly most abundant within the Canadian fauna, it slightly overlaps into the Alleghanian during the breeding season, though not common therein.

County Records.—Androscoggin, "fairly common summer resident" (Call); Aroostook, "occurs at Fort Fairfield" (Batchelder, Bull. Nutt. Orn. Club, Vol. 7, p. 109); Cumberland, "common summer resident" (Brown's Cat. Birds of Portland, p. 7); "not rare migrant" (Mead); Franklin, "common summer resident" (Swain); Kennebec, "common summer resident" (Gardiner Branch); Knox, "summer" (Rackliff); Oxford, "breeds commonly" (Nash); Penobscot, "quite common summer resident" (Knight); Piscataquis, "common, breeds" (Homer); Sagadahoc, (Spratt); Somerset, "quite common summer resident" (Morrell); Waldo, "summer resident" (Spratt); Washington, "abundant summer resident" (Boardman).

278. (659). Dendroica pensylvanica (*Linn.*). Chestnut-sided Warbler.

Quite common as a summer resident of very general distribution.

County Records.—Androscoggin, "common summer resident" (Johnson); Aroostook, "common at Fort Fairfield," (Batchelder, Bull. Nutt. Orn. Club, Vol. 7, p. 109); Cumberland, "common summer resident" (Mead); Franklin, "common summer resident" (Swain); Kennebec, "rare summer resident" (Gardiner Branch); Knox, "summer" (Rackliff); Oxford, "breeds commonly" (Nash); Penobscot, "common summer resident" (Knight); Piscataquis, "common, breeds" (Homer); Sagadahoc, "rare, two spring specimens" (Spinney); Somerset, "common summer resident" (Morrell); Waldo, (Spratt); Washington, "not uncommon summer resident" (Boardman); York, "quite common" (Adams).

279. (660). Dendroica castanea (*Wils.*). Bay-breasted Warbler.

As a summer resident this species is strictly confined to the Canadian fauna, and here it is not at all common except locally. As a migrant it probably occurs throughout the state, but it may be ranked as of rare occurrence in most places.

County Records.—Androscoggin, "rare migrant" (Johnson); Cumberland, "rare" (Mead); Franklin, "rare migrant" (Richards); Knox, "rare in summer" (Rackliff); Oxford, "breeds" (Maynard's List of Birds of Coos County, N. H., and Oxford County, Me., p. 9); Penobscot, "very rare even as a migrant, a nest and one egg taken near Orono, are in the University of Maine collection" (Knight); Piscataquis, "migrant, not

uncommon" (Homer) ; Sagadahoc, "one specimen iu spring" (Spinney) ;
Somerset, "rare migrant" (Morrell) ; Washington, "not uncommon summer resident" (Boardman).

280. (661). Dendroica striata (*Forst.*). Black-poll Warbler.

Quite common in migrations, also quite abundant as a summer
resident of the Canadian fauna.

County Records.—Androscoggin, "fairly common migrant" (Johnson) ;
Cumberland, "common transient near Portland" (Brown's Cat. Birds of
Portland, p. 8) ; Franklin, "rare summer resident" (Swain) ; Kennebec,
"migrant" (Larrabee) ; Knox, "often common" (Rackliff) ; Oxford,
"migrant" (Nash) ; Penobscot, "quite common migrant, have seen it in
late May" (Knight) ; Piscataquis, "common migrant" (Homer) ; Sagadahoc, "common migrant" (Spinney) ; Somerset, "rare migrant" (Morrell) ;
Washington, "not uncommon summer resident" (Boardman).

281. (662). Dendroica blackburniæ. (*Gmel.*). Blackburnian Warbler.

Usually quite rare but often locally abundant, both as a summer
resident and also in migrations.

County Records.—Androscoggin, "rare summer resident" (Johnson) ;
Aroostook, "seldom seen at Fort Fairfield" (Batchelder, Bull. Nutt. Orn.
Club, Vol. 7, p. 109) ; Cumberland, "not very common summer resident"
(Brown's Cat. Birds of Portland, p. 8), "common migrant" (Mead) ;
Franklin, "rare summer resident" (Swain) ; Hancock, "summer resident"
(Murch) ; Kennebec, "very rare summer resident" (Powers) ; Knox,
"rare migrant" (Rackliff) ; Oxford, "breeds very rarely" (Nash) ; Penob-
scot, "some years quite common in migrations while other seasons it
is not observed, rare as a summer resident" (Knight) ; Piscataquis,
"common migrant" (Homer) Sagadahoc, "rare, two specimens in spring"
(Spinney) ; Somerset, "not uncommon summer resident" (Morrell) ;
Washington, "not uncommon summer resident" (Boardman).

282. (667). Dendroica virens (*Gmel.*). Black-throated Green
Warbler.

Very common in migrations, also a common summer resident in
most localities. In the summer it should be sought in the tops of
the taller evergreen trees and here it places its nest.

County Records.—Androscoggin, "fairly common summer resident"
(Johnson) ; Aroostook, "rather common at Fort Fairfield" (Batchelder,
Bull. Nutt. Orn. Club, Vol. 7, p. 109) ; Cumberland, "common summer
resident" (Mead) ; Franklin, "common summer resident" (Swain) ; Hancock, "summer resident" (Knight) ; Kennebec, "common summer resident" (Gardiner Branch) ; Knox, "summer" (Rackliff) ; Oxford, "breeds
commonly" (Nash) ; Penobscot, "common summer resident" (Knight) ;

Piscataquis, "common, breeds" (Homer); Sagadahoc, "common summer resident" (Spinney); Somerset, "quite common summer resident" (Morrell); Washington, "abundant summer resident" (Boardman); York, "migrant" (Adams).

283. (671). Dendroica vigorsii (*Aud.*). Pine Warbler.

The Pine-creeping Warbler is quite a common summer resident of most portions of the state, although being inclined to occur locally and, as its name would indicate, in pine forests.

County Records.—Androscoggin, "fairly common summer resident" (Johnson); Cumberland, "common summer resident, one nest" (Mead); Franklin. "rare migrant" (Richards); Kennebec, "very rare summer resident" (Powers); Oxford, "breeds rarely" (Nash); Penobscot, "quite rare summer resident" (Knight); Sagadahoc, "common migrant" (Spinney); Washington, "very rare, one specimen" (Boardman); York, "common migrant" (Adams).

284. (672 a). Dendroica palmarum hypochrysea *Ridgw.* Yellow Palm Warbler.

One of our commonest Warblers during the migrations. Its chestnut poll, yellow breast streaked with dusky, flycatcher-like actions, and its constant habit of twitching its tail render it one of the most easily identified of our Warblers. It is one of the first Warblers to arrive from the south, being on hand by the last of April or sometimes as early as the 20th, while in the fall it does not depart till late October. The fact that this species nests on the ground makes its nest comparatively difficult to discover, and it is only within the past four years that it has been ascertained to be quite a common breeder of local distribution, within a few counties of the Canadian fauna. Bangor is the southernmost locality where it occurs at all commonly in the breeding season, and here its abundance is limited to the precincts of a large juniper bog, locally known as Orono Bog. Mr. Chas. H. Whitman of Bangor first found a nest with young of this bird at the above locality. Since then I have had the pleasure of taking two sets of their eggs, both in early June, and examining a number of nests with young. I have found nests with young on Memorial Day. (For records Cf. Knight, The Oologist, February, 1893, p. 54, and The Nidologist, June, 1895, p. 140). The number of birds breeding here varies from year to year, but even when they are rarest a person cannot walk one-fourth mile in this bog without seeing one or more of them. As it is nearly seven miles long, though interrupted by

occasional high land, and averages half a mile in width, the number of birds breeding there must be considerable. A set of eggs
with the parent bird was taken near Pittsfield, on June 13th, 1894,
by Mr. H. H. Johnson. (Cf. Johnson, The Nidologist, June,
1895, p. 140). The late Mr. Anson Allen of Orono is said to have
taken a nest and eggs near that place some years ago, but I am not
aware that they were ever recorded. Wherever open juniper bogs
occur, within the Canadian fauna, this bird may confidently be
expected to occur as a summer resident, although it is not exclusively confined to boggy localities.

County Records.—Androscoggin, "common migrant" (Johnson); Cumberland, "common migrant" (Mead); Franklin, "rare migrant" (Richards); Hancock, "occurs, status unknown" (Knight); Kennebec, "rare
migrant" (Powers); Knox, "transient" (Norton); Oxford, "fairly common migrant" (Johnson); Penobscot, "locally common summer resident" (Knight); Sagadahoc, (Spratt); Somerset, "common migrant,
rare summer resident" (Morrell); Washington, "very abundant summer
resident" (Boardman); York, "vernal migrant" (Adams).

Genus SEIURUS Swainson.

285. (674). Seiurus aurocapillus (*Linn.*). Oven-bird.

A common summer resident of general occurrence, resorting to
the solitudes of the woods. Here its song of "Teacher, teacher,
teacher" may be heard during the nesting season. Its roofed,
oven-like nest is placed on the ground, usually at the base of
some small shrub, and is difficult to find unless the parent bird is
flushed from it.

County Records.—Androscoggin, "fairly common summer resident"
(Johnson); Aroostook, "rather common at Fort Fairfield" (Batchelder,
Bull. Nutt. Orn. Club, Vol. 7, p. 110); Cumberland, "common summer
resident" (Mead); Franklin, "common summer resident" (Swain); Hancock, "summer resident" (Knight); Kennebec, "quite common summer
resident" (Powers); Knox, "summer" (Racklift); Oxford, "common,
breeds" (Nash); Penobscot, "quite common summer resident" (Knight);
Piscataquis, "common summer resident" (Homer); Sagadahoc, "common
summer resident" (Spinney); Somerset, "common summer resident"
(Morrell); Waldo, (Spratt); Washington, "very abundant summer
resident" (Boardman); York, "quite common summer resident" (Adams).

286. (675). Seiurus noveboracensis (*Gmel.*). Water-Thrush.

Migrant in southern Maine, while in the counties of the Canadian
fauna it occurs as a summer resident of somewhat local distribution and variable abundance.

County Records.—Androscoggin, "rare summer resident" (Call); Aroostook, "breeding at Fort Fairfield" (Batchelder, Bull. Nutt. Orn. Club, Vol. 7, p. 110); Cumberland, "quite rare" (Mead); Franklin, (Lee & McLain); Kennebec, "very rare summer resident" (Dill); Knox, "transient" (Norton); Oxford, "breeds" (Maynard's List of Birds of Coos Co., N. H., and Oxford Co., Me., p. 3); Penobscot, "rare summer resident" (Knight); Piscataquis, "common summer resident" (Homer); Sagadahoc, "not common migrant" (Spinney); Somerset, "common summer resident" (Morrell); Waldo, "rare" (Spratt); Washington, "common summer resident" (Boardman).

287. (676). Sciurus motacilla (*Vieill.*). Louisiana Water-Thrush.

Of accidental occurrence as a straggler from the south. Specimens taken at Norway, Oxford County, in 1865, by Mr. Irving Frost, and at Waterville, Kennebec County, by Prof. Hamlin during the same year, are recorded by Stearns. (Cf. Stearns' "New England Bird Life," p. 159).

Genus GEOTHLYPIS Cabanis.

Subgenus OPORORNIS Baird.

288. (678). Geothlypis agilis (*Wils.*). Connecticut Warbler,

The records of this bird are very meager, but it probably occurs as a regular though rare fall migrant.

County Records.—Cumberland, "one taken August 30, 1878" (Brown's Cat. Birds of Portland, p. 9), "one taken at Westbrook, September 20, 1896" (Norton); York, "one at Saco in September, 1885, one September 8th and another September 15, 1886" (Cf. Goodale, The Auk, Vol. 4, p. 77).

Subgenus GEOTHLYPIS Cabanis.

289. (679). Geothlypis philadelphia (*Wils.*). Mourning Warbler.

Occurs as a rare transient in the southern counties, and a rare summer resident of the Canadian fauna.

County Records.—Androscoggin, "rare migrant" (Johnson); Aroostook, "common at Fort Fairfield" (Batchelder, Bull. Nutt. Orn. Club, Vol. 7, p. 110); Cumberland, "rare" (Mead); Franklin, "rare summer resident" (Swain); Kennebec, (Royal); Oxford, "occurs at Upton" (Brewster, Bull. Nutt. Orn. Club, Vol. 3, p. 61); Sagadahoc, "not uncommon migrant" (Spinney); Washington, "very rare" (Boardman).

290.   (681).   Geothlypis trichas (*Linn.*).   Maryland Yellow-throat.

A common summer resident, frequenting grassy, bush-interspersed meadows, low, bushy clearings and similar localities.   Its song resembles the syllables "peachity, peachity, peachity," while the alarm note is a harsh "chit." It prefers to skip about in the low bushes, keeping fairly well concealed from observation, but always making its presence known by uttering its alarm note when its precincts are intruded upon.

County Records.—Androscoggin, "common summer resident" (Johnson); Aroostook, "Fort Fairfield, common" (Batchelder, Bull. Nutt. Orn. Club, Vol. 7, p. 110); Cumberland, "common summer resident" (Mead); Franklin, "common summer resident" (Swain); Hancock, "common summer resident, noted on Deer Isle" (Knight); Kennebec, "quite common summer resident" (Powers); Knox, "summer" (Rackliff); Oxford, "breeds commonly" (Nash); Penobscot, "common summer resident" (Knight); Piscataquis, "common summer resident" (Homer); Sagadahoc, "common summer resident" (Spinney); Somerset, "common summer resident" (Morrell); Waldo, (Spratt); Washington; "abundant summer resident" (Boardman); York, "common summer resident" (Adams).

Genus ICTERIA Vieillot.

291.   (683).   Icteria virens (*Linn.*).   Yellow-breasted Chat.

An accidental visitor from the south of which only three specimens have come to my knowledge. Two of these are Cumberland County specimens, one a male, being taken at North Bridgton, June 6th, 1880, by Mr. J. C. Mead, while the other is from Portland, being recorded by Brown. (Cf. Brown, The Auk, Vol. 11, p. 331). A specimen taken at Elliot, York County, is given in Smith's List of the Birds of Maine.

Genus SYLVANIA Nuttall.

292.   (685).   Sylvania pusilla (*Wils.*).   Wilson's Warbler.

A rare summer resident of the Canadian fauna, rare as a migrant elsewhere in the state.

County Records.—Androscoggin, "sometimes rear their young in this county" (Walter's "Birds of Androscoggin County, p. 9); Aroostook, "breeds at Fort Fairfield" (Batchelder, Bull. Nutt. Orn. Club, Vol. 7, p. 110); Cumberland, "uncommon transient" (Brown's Cat. Birds of Portland, p. 9), "I have record of its occurrence in Westbrook, September

11, 1895" (Norton); Franklin, "rare migrant" (Swain); Kennebec. (Robbins); Penobscot, "rare" (Knight); Piscataquis, "rare summer resident" (Whitman); Sagadahoc, "not uncommon migrant" (Spinney); Somerset, "rare summer resident" (Morrell); Washington, "not common summer resident" (Boardman).

293. (686). Sylvania canadensis (*Linn.*). Canadian Warbler.

A quite common summer resident of general occurrence, but more generally distributed as such within the Canadian fauna.

County Records.—Androscoggin, "rare summer resident" (Johnson); Aroostook, "common at Fort Fairfield" (Batchelder, Bull. Nutt. Orn. Club, Vol. 3, p. 61); Cumberland, "rather common summer resident" (Brown's Cat. Birds of Portland, p. 10); Franklin, "common summer resident" (Richards); Kennebec, "rare summer resident" (Larrabee); Knox, "summer" (Racklifl); Oxford, "occurs at Upton" (Brewster, Bull. Nutt. Orn. Club, Vol. 3, p. 61); Penobscot, "quite common summer resident" (Knight); Piscataquis. "not common" (Homer); Sagadahoc, "not uncommon migrant" (Spinney); Somerset, "quite common summer resident" (Morrell); Washington, "common summer resident" (Boardman).

Genus SETOPHAGA Swainson.

294. (687). Setophaga ruticilla (*Linn.*). American Redstart.

Common summer resident everywhere, irrespective of faunal areas, within the state.

County Records.—Androscoggin, "common summer resident" (Johnson); Aroostook, "common" (Batchelder. Bull. Nutt. Orn. Club, Vol. 7, p. 110); Cumberland, "common summer resident" (Mead); Franklin, "common summer resident" (Swain); Hancock, "summer resident" (Murch); Kennebec, "abundant summer resident" (Sanborn); Knox, "summer" (Rackliff); Oxford, "common" (Nash); Penobscot, "very common summer resident" (Knight); Piscataquis, "common summer resident" (Homer); Sagadahoc, "common summer resident" (Spinney); Somerset, "common summer resident" (Morrell); Waldo, "rare" (Spratt); Washington, "very abundant summer resident" (Boardman).

Genus ANTHUS Bechstein.

Subgenus ANTHUS.

295. (697). Anthus pensilvanicus (*Lath.*). American Pipit.

A migrant of irregular abundance, often common in autumn and usually rare in spring.

County Records.—Androscoggin, "fall migrant" (Walter's Birds of
Androscoggin County, p. 6); Cumberland, "irregularly abundant in the
inland towns in autumn, a flock of two dozen was observed in Westbrook,
May 15, 1889" (Norton); Franklin, "not common" (Lee & McLain);
Knox, "migrant" (Rackliff); Penobscot, "common some falls, rare
others" (Knight); Piscataquis, "common migrant" (Homer); Somerset,
"common migrant" (Morrell); Washington, "common migrant" (Boardman).

Family TROGLODYTIDÆ.   Wrens, Thrashers, etc.

Subfamily MIMINÆ.   Thrashers.

Genus MIMUS Boie.

296.   (703).   Mimus polyglottos (*Linn.*).   Mockingbird.

Although many Mockingbirds have been taken in the state, still
it seems very evident that they must have all originally been cage-
birds which escaped from captivity.   While reliable observers have
reported seeing these birds at liberty, and, even in midwinter,
observed the same individuals for many successive days or weeks,
still the very fact that such a southern bird should be here in winter
shows that its presence is due primarily to human agency.   While
specimens have been taken which show no signs of ever being in
captivity, still we would not expect such indications of former
days of captivity to persist in case they had been free for several
weeks.   In The Auk for April, 1897, Mr. N. C. Brown records a
specimen which was seen at Portland, January 19, and at intervals
until February 15, while four days later one of Mr. Brown's neighbors saw it.   This was beyond a doubt an escaped cage-bird,
although Mr. Brown states that it showed no evidences of former
captivity.   Such evidences would be difficult to detect in a living
bird at some distance from the observer.   I have been at great
difficulty to detect proofs of former captivity in skins of escaped
birds when actually in my hands.

County Records.—Cumberland, "have one, an escaped cage bird, taken
at Gorham, August 12, 1890" (Norton), "one seen at Portland, January
19—February 19, 1897" (Cf. Brown, Auk, April 1897, p 225); Knox,
"one taken in February (Rackliff), "have one shot at Vinalhaven, February 1891, an escaped cage bird" (Norton); Oxford, (Nash); Piscataquis,
"one shot in Monson, October 20th, 1884, did not seem to have been a
caged bird" (Homer); Washington, "one observed near Calais in 1870"
(Boardman).

Genus GALEOSCOPTES Cabanis.

297. (704). Galeoscoptes carolinensis (*Linn.*). Catbird.

A summer resident of general distribution, but growing less abundant as the northern and eastern counties are approached.

County Records.—Androscoggin, "common summer resident" (Johnson); Aroostook, "very rare at Houlton, nests" (Batchelder, Bull. Nutt. Orn. Club, Vol. 7, p. 109); Cumberland, "common summer resident" (Mead); Franklin, "common summer resident" (Swain); Hancock, "summer resident" (Murch); Kennebec, "quite common summer resident" (Gardiner Branch); Knox, "summer" (Racklift); Oxford, "breeds" (Nash); Penobscot, "quite common summer resident but far less abundant than in former years" (Knight); Piscataquis, "summer resident, not common" (Homer); Sagadahoc, "common summer resident" (Spinney); Somerset, "common summer resident" (Morrell); Waldo, (Spratt); Washington, "not very abundant summer resident" (Boardman); York, "common" (Adams).

Genus HARPORHYNCHUS Cabanis.

Subgenus METHRIOPTERUS Reichenbach.

298. (705). Harporhynchus rufus (*Linn.*). Brown Thrasher.

One of the best test species of the Alleghanian fauna which we have. A common summer resident within its limits while elsewhere it is of rare or casual occurrence.

County Records.—Androscoggin, "common summer resident" (Johnson); Cumberland, "common in some parts of the county but not so near Portland" (Brown's Cat. Birds of Portland, p. 4), "rare summer resident" (Mead); Franklin, "rare summer resident" (Richards); Kennebec, "common summer resident" (Gardiner Branch); Knox, "summer" (Racklift); Oxford, "breeds" (Nash); Sagadahoc, "common summer resident" (Spinney); York, "quite common summer resident" (Adams).

Subfamily TROGLODYTINÆ. Wrens.

Genus THRYOTHORUS Vieillot.

Subgenus THRYOTHORUS.

299. (718). Thryothorus ludovicianus (*Lath.*). Carolina Wren.

There is in the collection of birds made by Prof. Chas. Hamlin, and at present the property of Colby University, a specimen of this bird taken at Waterville. Upon this evidence we may admit the species to the list as accidental. Mr. J. Waldo Nash of Norway, Me., writes me that he has seen two of these birds there, but unfortunately he did not procure either of them so as to positively verify the record.

Genus TROGLODYTES Vieillot.

Subgenus TROGLODYTES.

300. (721). Troglodytes aëdon *Vieill.* House Wren.

Formerly quite common in many places where it does not now occur. Being a typical bird of the Alleghanian fauna, we need only look for it within these limits, and here it was formerly locally abundant. It was formerly common in Penobscot County, near Bangor, but has not been observed there for ten years. In other portions of the state this species seems likewise to be contracting its northern range.

County Records.—Androscoggin, "tolerably common summer resident" (Call); Cumberland, "not seen in many years, formerly occurred" (Mead); Franklin, "rare summer resident" (Richards); Hancock, "some years ago a pair built in a bird house in my yard" (Dorr); Kennebec (Given in Hamlin's List of the Birds of Waterville, Report of Maine Board of Agriculture, 1865, pp. 168-173); Knox, "formerly occasional visitant" (Norton); Oxford, "breeds commonly" (Nash); Penobscot, "formerly nested in Bangor, not reported for ten years" (Knight); Somerset "rare summer resident" (Morrell).

Subgenus ANORTHURA Rennie.

301. (722). Troglodytes hiemalis *Vieill.* Winter Wren.

A good test species of the Canadian fauna, and quite a common summer resident within its limits, while elsewhere it is a common migrant. A few individuals may remain through winter in the southern counties.

County Records.—Androscoggin, "fairly common migrant" (Johnson); Aroostook, "breeds at Houlton" (Batchelder, Bull. Nutt. Orn. Club, Vol. 7, p. 109); Cumberland, "rare migrant" (Mead); Franklin, "rare summer resident" (Richards); Hancock, "summer resident" (Knight); Kennebec, "rare migrant" (Larrabee); Knox, "winter" (Rackliff); Oxford, "breeds commonly" (Nash); Penobscot, "quite common summer resident" (Knight); Piscataquis, "common summer resident" (Homer); Somerset, "quite common summer resident" (Morrell); Waldo, "summer resident" (Spratt); Washington, "summer resident, not abundant" (Boardman); York, "migrant" (Adams).

Family CERTHIIDÆ.   Creepers.

Genus CERTHIA Linnæus.

302.   (726).   Certhia familiaris americana   (*Bonap.*).   Brown Creeper.

Common in migrations, quite common summer resident of the Canadian fauna, and some winters it also occurs.

County Records. — Androscoggin, "common migrant" (Johnson); Aroostook, "breeds at Houlton and Fort Fairfield" (Batchelder, Bull. Nutt. Orn. Club, Vol. 7, p. 109); Cumberland, "common migrant" (Mead); Franklin, "rare summer resident" (Swain); Kennebec, "quite common resident" (Powers); Knox, "migrant" (Racklift); Oxford, "common, breeds" (Nash); Penobscot, "summer resident and have seen it also in February" (Knight); Piscataquis, "some years resident" (Homer); Sagadahoc, "common migrant" (Spinney); Somerset, "common resident" (Morrell); Waldo, (Spratt); Washington, "not very common, breeds" (Boardman); York, fairly common vernal migrant" (Adams).

Family PARIDÆ.   Nuthatches and Tits.

Subfamily SITTINÆ.   Nuthatches.

Genus SITTA Linnæus.

303.   (727).   Sitta carolinensis *Lath.*   White-breasted Nuthatch.

Common and of general occurrence in migrations, also less abundant as a permanent resident.

County Records.—Androscoggin, "fairly common resident" (Call); Aroostook, "occurs at Houlton" (Batchelder, Bull. Nutt. Orn. Club, Vol. 7, p. 109); Cumberland, "common resident" (Mead); "uncommon in migrations and winter" (Brown's Cat. Birds of Portland, p. 5); Franklin, "common resident" (Swain); Hancock, "summer resident" (Murch); Kennebec, "quite common resident" (Gardiner Branch); Knox, "rare migrant" (Racklift); Oxford, "breeds" (Nash); Penobscot, "very common migrant, rare in summer and exceedingly so in winter" (Knight); Piscataquis, "common resident" (Homer); Sagadahoc, (Spratt); Somerset, "not common resident" (Morrell); Waldo, (Spratt); Washington, "rare, breeds" (Boardman); York, "not very common" (Adams).

304.   (728).   Sitta canadensis *Linn.*   Red-breasted Nuthatch.

Resident within the Canadian fauna, slightly more abundant in summer.   Elsewhere chiefly occurs as a migrant and winter resident, common.

County Records.—Androscoggin, "rare summer resident" (Johnson);
Aroostook, (Batchelder, Bull. Nutt. Orn. Club, Vol. 7, p. 109); Cumberland, "common winter migrant" (Mead); Franklin, "common resident"
(Richards); Hancock, "breeds on the wooded islands of Penobscot Bay,
and also common inland" (Knight); Kennebec, "quite common summer
resident" (Gardiner Branch); Knox, "resident" (Rackliff); Oxford,
"common resident, breeds" (Nash); Penobscot, "resident, breeds commonly" (Knight); Piscataquis, "common summer resident" (Homer);
Sagadahoc, "common winter resident, one pair remained one summer"
(Spinney); Somerset, "common, not common summer resident" (Morrell); Waldo, (Spratt); Washington, "abundant, breeds" (Boardman);
York, "quite common resident" (Adams).

Subfamily PARINÆ.   Titmice.
Genus PARUS Linnæus.
Subgenus PARUS Linnæus.

305.   (735.)   Parus atricapillus *Linn.*   Chickadee.

One of our commonest resident species, occurring everywhere.

This confiding, curiosity loving bird pronounces its name very
plainly upon every occasion, so there is no need of its going uniden-
tified, even by the merest tyro.   I have often while in the woods
drawn a small band of these birds to within a few feet of me by
imitating their call, or by making a whistling or squeaking noise.
In addition to their cry of "chick-a-dee-dee" they utter a great
variety of chirps and whistles.   In the spring time their mating
call is a sweet, whistled succession of two or three notes which
cannot readily be put on paper.

County Records.—Androscoggin, "abundant resident" (Johnson);
Aroostook, "Fort Fairfield and Houlton" (Batchelder, Bull. Nutt. Orn.
Club, Vol. 7, p. 109); Cumberland, "common resident" (Mead); Franklin, common resident" (Swain); Hancock, "common resident, especially
common on the wooded islands" (Knight); Kennebec, "abundant resident" (Gardiner Branch); Knox, "resident" (Rackliff); Oxford, "breeds
commonly" (Nash); Penobscot, "common resident" (Knight); Piscataquis, "common resident" (Homer); Sagadahoc, "common resident"
(Spinney); Somerset, "common resident" (Morrell); Waldo, "common
resident" (Knight); Washington, "common resident" (Boardman);
York, "common resident" (Adams).

306.   (740).   Parus hudsonicus *Forst.*   Hudsonian Chickadee.

Chiefly occurring as a somewhat rare winter visitor.   It also is
resident in the extreme northern and eastern counties, though very

rare in this guise. It has been reported by Dr. Brewer as being seen on Mt. Desert in summer, July and August, this being the southernmost record for summer.

County Records.—Androscoggin, "rare winter visitor" (Johnson); Cumberland, "rare" (Mead); Franklin, "rare winter resident" (Richards); Hancock, "winter migrant" (Murch); Oxford, "occurs at Upton" (Brewster, Bull. Nutt. Orn. Club, Vol. 3, p. 20); Penobscot, "winter visitor of variable abundance, usually rare" (Knight); Piscataquis, "common winter visitor" (Homer); Washington, "not common, a few breed" (Boardman).

Family SYLVIIDÆ. Warblers, Kinglets, Gnatcatchers.
Subfamily REGULINÆ. Kinglets.
Genus REGULUS Cuvier.

307. (748). Regulus satrapa *Licht.* Golden-crowned Kinglet.

Of general distribution throughout our limits and resident to a certain extent. While most abundant in fall and spring, these birds are not uncommon in winter, being usually seen in flocks associated with Nuthatches, Chickadees and Creepers. In the summer they resort to the topmost branches of the taller evergreens, and here the nest is usually situated.

County Records.—Androscoggin, "common winter resident" (Johnson); Cumberland, "properly a summer resident, suspect a few remain through winter, abundant in migrations" (Brown's Cat. Birds of Portland, p. 5); "common winter migrant" (Mead); Franklin, "common migrant" (Swain); Hancock, "have seen it in summer" (Knight); Kennebec, "very rare winter resident" (Powers); Knox, "resident" (Rackliff); Oxford, "common" (Nash); Penobscot, "common in summer, breeds, abundant in migrations, rare in winter" (Knight); Piscataquis, "resident" (Whitman); Sagadahoc, "migrant, common in fall of 1896" (Spinney); Somerset, "common, a frequent summer resident" (Morrell); Waldo, "summer resident" (Spratt); Washington, "quite common, a few winter, rarely breeds" (Boardman); York, "not common migrant" (Adams).

308. (749). Regulus calendula (*Linn.*). Ruby-crowned Kinglet.

Not so common as the preceeding, and more likely to be seen in the migrations, than at any other season. Winters south of the state and in summer the majority pass north of our boundaries. However it is an indisputable fact that a limited number occasionally remain here through the summer. On May 31, 1897, while collecting in a thick woods of mixed spruce and fir, my attention was attracted by the constantly recurring song of a Kinglet. With the aid of a pair of opera glasses I located the songster, and found that he was accompanied by his mate who was engaged in building her nest. She would seek suitable material in the immediate vicinity, and with her mouth filled with huge pieces of moss, gleaned from the tree trunks, she would repair to the top of a spruce tree which was near at hand. I climbed the tree and located the nest near the extremity of a limb, 25 feet from the ground, but well concealed from observation from below. It was then a mere foundation of mosses and had evidently just been commenced. The birds were somewhat shy, but by careful observation I failed to detect the yellow crown patch of *satrapa*, nor was I able to fully satisfy myself that they were *calendula*. I made a number of subsequent visits to the nest and watched the actions of the birds, becoming fully satisfied of their identity. On June 15th the nest was ready to be lined, and I did not again visit it until the 24th, when I found it was deserted, this doubtless being due to my unusually close examination of it during my previous visit.

The nest, which is now in my possession, was situated near the end of a limb, 25 feet from the ground and about 8 feet from the main trunk. It was supported by a number of small twigs which drooped from the limb and was directly under it. Exteriorly it is composed of mosses, mostly such species as grow on the trunks of trees, mixed with a few lichens of the genus *Cladonia*, *Parmelia*, and *Usnea*. Viewed from a distance of a few feet it looked like a green ball of moss. Interiorly it is composed of *Usnea longissima*, closely interwoven and intimately mixed with feathers and small quantities of moss. The lining is not completed. The exterior depth is four and the interior three inches, while the outside diameter is three and the inside one and one-half inches. A few days later I visited the same woods, obtained a glimpse of the birds, and heard the song of the male, but soon lost sight of them. The locality was about four miles from Orono, Penobscot County.

County Records.—Androscoggin, "fairly common" (Johnson); Cumberland, "common transient" (Brown's Cat. Birds of Portland, p. 5); Franklin, "common migrant" (Richards); Hancock, "occurs, status not known" (Knight); Kennebec, "very rare migrant" (Dill); Knox, "rare migrant" (Racklifl); Oxford, "rare" (Nash); Penobscot, "common migrant, rare summer resident"(Knight); Piscataquis,"common migrant" (Homer); Sagadahoc, "rare. one specimen" (Spinney); Somerset, "rare, have once seen it when I was sure it was breeding" (Morrell); Washington, "rare, may breed" (Boardman); York, "not common migrant" (Adams).

Subfamily POLIOPTILINÆ. Gnatcatchers.

Genus POLIOPTILA Sclater.

309. (751)., Polioptila cærulea (*Linn.*). Blue-gray Gnatcatcher.

An accidental estray from the south, only two positive instances of its occurrence being known. These both rest on the excellent authority of Mr. Brown. One of these was observed at Cape Elizabeth, Cumberland County, August 29, 1880. (Brown's Cat. Birds of Portland, p. 5). The second example of this species was observed at the same place, April 18, 1896. (Cf. Brown, The Auk, Vol. 13, p. 264).

Family TURDIDÆ. Thrushes, Solitaires, Stonechats, Bluebirds, etc.

Subfamily TURDINÆ. Thrushes.

Genus TURDUS Linnæus.

Subgenus HYLOCICHLA Baird.

310. (755). Turdus mustelinus *Gmel.* Wood Thrush.

Of rare occurrence as a summer resident, being found only near the southern and southwestern boundaries.

County Records.—Franklin, "rare summer resident, have taken nest, eggs and bird" (Swain); Kennebec, "given in Smith's List as having been taken at Vassalboro" (Cf. Smith's List of the Birds of Maine, Forest and Stream); Oxford, "have secured two sets of eggs during a period of eight years" (Nash); York, "taken at Saco" (Goodale, The Auk, Vol. 2, p. 215).

311. (756). Turdus fuscescens *Steph.* Wilson's Thrush.

A common summer resident, most abundant in the counties of the Alleghanian fauna, although not uncommon in the extreme northern and eastern counties.

County Records.—Androscoggin, "common summer resident" (Johnson); Aroostook, "breeds at Houlton" (Batchelder, Bull. Nutt. Orn. club, Vol. 7, p. 108); Cumberland, "common summer resident" (Brown's Cat. Birds of Portland, p. 3), "rare" (Mead); Franklin, "common summer resident" (Swain); Hancock, "common summer resident" (Dorr); Kennebec, "common summer resident" (Gardiner Branch); Oxford, "breeds commonly" (Nash); Penobscot, "quite common summer resident" (Knight); Sagadahoc, (Spratt); Somerset, "common summer resident" (Morrell); Waldo, (Spratt); Washington, "not uncommon summer resident" (Boardman); York, "common summer resident" (Adams).

312. (757). Turdus aliciæ *Baird*. Gray-cheeked Thrush.

Occurs within our limits as a migrant only, breeding north of the United States. While it must occur in considerable numbers during the migrations, the fact remains that it has escaped the notice of nearly all the collectors of the state, doubtless owing to the resemblance to its near relative, the Olive-backed Thrush.

County Records.—Cumberland, "uncommon transient" (Brown's Cat. Birds of Portland, p. 3).

313. (758a). Turdus ustulatus swainsonii (*Cab.*). Olivebacked Thrush.

A somewhat rare summer resident in many parts of the state, while in some places it is common. During the nesting season it need not be looked for except within the Canadian fauna, while elsewhere it occurs in the migrations.

County Records.—Androscoggin, "rare summer resident" (Johnson); Aroostook, "Fort Fairfield and Houlton" (Batchelder, Bull. Nutt. Orn. Club, Vol. 7, p. 108); Cumberland, "rare migrant" (Mead); Franklin, "rare summer resident" (Richards); Hancock, "summer resident" (Murch); Kennebec, "very rare summer resident" (Robbins); Knox, "summer" (Rackliff); Oxford, "breeds rarely" (Nash); Penobscot, "not common summer resident" (Knight); Piscataquis, "common summer resident" (Homer); Washington, "not common summer resident" (Boardman).

314. (759b). Turdus aonalaschkæ pallasii (*Cab.*). Hermit Thrush.

Our commonest Thrush as a migrant and summer resident. I have seen it in late October and also early April.

County Records.—Androscoggin, "fairly common summer resident" (Johnson); Cumberland, "common summer resident" (Brown's Cat. Birds of Portland, p. 3), "common summer resident" (Mead); Franklin, "common summer resident" (Swain); Hancock, "summer resident"

(Murch); Kennebec, "common summer resident" (Gardiner Branch); Knox, "summer" (Rackliff); Oxford, "breeds commonly" (Nash); Penobscot, "common summer resident and breeder" (Knight); Piscataquis, "common summer resident" (Homer); Sagadahoc, "common summer resident" (Spinney); Somerset, "not very common summer resident" (Morrell); Waldo, "common summer resident" (Knight); Washington, "abundant summer resident" (Boardman); York, (Adams).

Genus MERULA Leach.

315. (761). Merula migratoria (*Linn.*). American Robin.

An abundant summer resident everywhere. It has also been observed in winter near our southern boundary, but probably does not regularly remain throughout the entire season.

County Records.—Androscoggin, "abundant summer resident" (Johnson); Aroostook, "seen at Fairfield" (Batchelder, Bull. Nutt. Orn. Club, Vol. 7, p. 108); Cumberland, "common summer resident" (Mead); Franklin, "common summer resident" (Swain); Hancock, "common summer resident" (Murch); Kennebec, "abundant summer resident" (Gardiner Branch); Knox, "summer resident" (Rackliff); Oxford, "breeds commonly" (Nash); Penobscot, "abundant summer resident, April to November" (Knight); Piscataquis, "common summer resident" (Homer); Sagadahoc, "common summer resident" (Spinney); Somerset, "common summer resident" (Morrell); Waldo, "common summer resident" (Knight); Washington, "abundant summer resident" (Boardman); York, "formerly abundant summer resident, wintered in 1888 and 1889, only two nests seen in 1896" (Adams).

Genus SIALIA Swainson.

316. (766). Sialia sialis (*Linn.*). Bluebird.

Formerly of common and general occurrence as a summer resident. During the past two years, 1895 and 1896, the species has been of very rare occurrence in most places, although locally common in a few places.

County Records.—Androscoggin, "common summer resident" (Johnson); Aroostook, "breeding at Houlton" (Batchelder, Bull. Nutt. Orn. Club, Vol. 7, p. 109); Cumberland, "common (?) summer resident" (Mead); Franklin, "summer resident, not common at present" (Swain) Hancock, "rare summer resident" (Murch); Kennebec, "very rare summer resident" (Powers); Knox, "summer" (Rackliff); Oxford, "breeds" (Nash); Penobscot, "formerly common, now rare, only about fifteen seen in the past two years" (Knight); Piscataquis, "common summer resident" (Homer); Sagadahoc, "common summer resident" (Spinney); Somerset, "formerly a common summer resident, for the past two years a rare migrant" (Morrell); Waldo, (Spratt); Washington, "not common summer resident" (Boardman); York, "rare visitant, formerly common" (Adams).

INTRODUCED SPECIES.

This list contains such birds as have been introduced into the state, never having naturally occurred here, and which have been known to breed after their liberation. Such species as the Prairie Hen, Sharp-tailed Grouse, Capercailzie, Black Grouse and European Quail, which were either let loose in limited numbers and known not to have survived, or whose survival and breeding is in doubt, are not considered entitled to a place in this list.

317.   Columba livia.   Domestic Pigeon.

Although introduced by man and originally domesticated, this species has escaped and breeds abundantly in the cornices of dwellings, deserted lofts, bridges and similar situations within our larger towns and cities. It is therefor seemingly entitled to a place in this list. These birds are not at all particular regarding the time of the year they choose for nesting. This present winter, 1897, I have, during the month of February, seen Pigeons engaged in incubating their eggs in nests in the cornices of houses, exposed to the force of every storm. One such nest I passed every day on my way down town. In late January I also saw a young bird, not long from the nest and still unable to fly very well. From their filthy habits they are very undesirable neighbors about houses and churches, though from this statement one should not infer that the birds themselves are dirty. It is the dirt they strew around that makes their presence undesirable.

318.   Passer domesticus (*Linn.*).   English Sparrow.

Although commonly known as English Sparrow because this bird was imported from England, the true name is European House Sparrow. It was first introduced to the state in 1854, when Col. William Rhodes liberated specimens at Portland. (Cf. Rhodes, Forest and Stream, Vol. 8, p. 165). Others were liberated in the same locality in 1858, by T. A. Dubois. (Cf. The English Sparrow in N. A., p. 18). Some individuals, who thought they were conferring a great benefit upon us, afterward liberated some of these birds at Bangor and Lewiston. From these centers of infection the curse has spread until the entire state is involved. Originally introduced in hopes they would exterminate injurious insects, the birds have swarmed over the entire state, although confined mainly to the vicinity of towns, villages and cities. The few insects they eat are mainly such as have been attracted to street lights, and would have perished without the interference of the Sparrows. Their food is chiefly vegetable matter, consisting in a great part of the undigested portions of grain in horse droppings, although they do not disdain bread, wheat and other delicacies which may be strewn by friendly hands for their benefit. Being resident, they take possession of the bird houses erected for the benefit of the Purple Martin and Tree Swallow while these latter are south for the winter, and upon the return of the rightful owners a fierce struggle for the nesting place results, although the Martins usually prove victorious. Were the Sparrows of any great benefit their presence could be endured, but they are not only useless, but noisy, quarrelsome, and often directly injurious. I have seen them in large flocks, feeding on oats and wheat standing in the fields near Bangor. Their huge filthy nests of straw and rubbish are placed in every conceivable situation, from the limb of a tree or the shelf of an electric street-light, to the cornice of a building or a hole in the eaves. When present in numbers, they paint and bespatter buildings with their filth. The very slight amount of good which they do by destroying injurious insects, would have been done far better and without half the bluster by the native birds that these foreigners have dispossessed. As regards distribution, we may safely say they are found in every town, village and city throughout the state. Recently while driving from Fort Fairfield to Limestone in Aroostook County, I was surprised to notice from one to three or more pair of these birds at nearly every country dwelling which I passed. They probably found abundance of food during the summer months, but when winter came they must either have been fed by the inmates of the farmhouses, or forced to retreat to the neighboring towns which were at least seven miles distant from some of the localities where the Sparrows were seen. I can personally vouch to having seen the pests in every county in the state.

HYPOTHETICAL LIST.

I have consigned to this list such species as probably occur in the state, but whose occurrence has not been proved by the actual capture of specimens within our limits. Here also are mentioned birds recorded from Maine or New England by previous authors, but which are now positively known not to have been taken within the actual limits of the state, although many have been taken very near our boundaries. Under each species its status or probable status will be found outlined in as comprehensive a manner as the information at my disposal will allow. Many birds taken at Grand Menan and other islands of that vicinity, which are politically a part of New Brunswick, have been given as Maine birds by previous writers on the subject. Such will be found discussed here to a greater or lesser extent.

Family URINATORIDÆ.   Loons.

Genus URINATOR Cuvier.

1. (9).   Urinator arcticus (*Linn.*).   Black-throated Loon.

This species is given in the A. O. U. Check List of North American Birds as being of casual occurrence in autumn and winter in the northern United States, east of the Rocky Mountains. In Smith's List of the Birds of Maine this species is given, but its occurrence does not seem to be satisfactorily shown. There is no doubt but what it will probably be added to our list upon the best of evidence before many years have gone by.

Family ALCIDÆ.   Auks, Murres, and Puffins.

Genus CEPPHUS Pallas.

2. (28).   Cepphus mandtii (*Licht.*).   Mandt's Guillemot.

The A. O. U. Check List gives this species as being found along the Atlantic coast, in winter as far south as Massachusetts. In view of this fact it is highly probable that it will ultimately be found to be a fairly regular winter visitant to our coast. On account of the resemblance of this species to the Black Guillemot it would easily pass unrecognized except by Ornithologists.

Family LARIDÆ. Gulls and Terns.

Genus GAVIA Boie.

3. (39). Gavia alba (*Gunn.*). Ivory Gull.

A specimen has been taken by Mr. Boardman at Grand Menan, New Brunswick, and hence it is accorded a place in this part of the list, although it has no claim as a bird of the state and is not likely to be taken within our limits.

Genus LARUS Linnæus.

4. (45). Larus kumlieni *Brewst.* Kumlien's Gull.

Probably a regular winter visitor to the state, and doubtless commoner than would seem possible from the few reports received concerning it. Mr. Harry Merrill of Bangor has a specimen of this bird, shot in the vicinity of Eastport and possibly in Maine waters, although very near to the New Brunswick limits. This species is also referred to in Smith's List of the Birds of Maine, in the Forest and Stream for April 12, 1883, under the White-winged Gull *of which Mr. Smith considers this bird a mere phase of plumage. In a recent letter from Mr. Smith he says: "I have examined a number of them in the flesh but have never shot them here, although I have observed them alive in Portland Harbor and at Scarboro." While this evidence is almost sufficient to give the bird a place in the list, still there seems to be some slight tinge of doubt that positively identified specimens have actually been taken in the state, and until this is proved beyond a doubt it will have to be assigned to the hypothetical list.

Family PROCELLARIIDÆ. Fulmars and Shearwaters.

Genus FULMARUS Stephens.

5. (86). Fulmarus glacialis (*Linn.*). Fulmar.

This is reported as a winter seabird at Grand Menan, by Mr. Boardman. It is also given in the A. O. U. List as occurring as far south as New Jersey. While there is no reasonable doubt but what it occurs as a winter visitor along the coast, yet there have been no specimens recorded from the state.

---

*Larus leucopterus *Faber*, or Iceland Gull.

6. (86 a). Fulmarus glacialis minor (*Kjœrbølling*). Lesser Fulmar.

Given in the A. O. U. List as occurring south to Massachusetts. It will probably be found associated with the preceding species along our coast, and it is only a question of time when both will be added to our list.

Genus PROCELLARIA Linnæus.

7. (104). Procellaria pelagica *Linn.* Stormy Petrel.

Given by Mr. Boardman as of accidental occurrence at Grand Menan. It is quite likely to occur accidentally along our coast.

Family FREGATIDÆ. Man-o'-War Birds.
Genus FREGATA Brisson.

8. (128). Fregata aquila *Linn.* Man-o'-War Bird.

Given in Smith's List of the Birds of Maine, Forest and Stream, Vol. 20, p. 185, as a straggler past the Maine coast to Nova Scotia, where it has been taken. It does not seem advisable to admit a species to the state on such presumptive evidence. Stearn's "New England Bird Life" p. 342, says: "Mr. Purdie's manuscript informs us that a specimen was taken, but not preserved, about twelve years ago, at Boothbay, Maine". This does not seem perfectly satisfactory grounds for admitting the species to the list.

Family ANATIDÆ. Ducks, Geese, and Swans.
Genus CAMPTOLAIMUS Gray.

9. (156). Camptolaimus labradorius (*Gmel.*). Labrador Duck.

This species probably formerly occurred along our coast in winter but has not been taken of late years. The last known example was taken at Grand Menan, N. B., in 1871, and Ornithologists now believe that the species is extinct. Mr. Boardman gives the species as formerly occurring at Grand Menan.

Genus CHEN Boie.

10. (169.1). Chen cærulescens (*Linn.*). Blue Goose.

This species will probably be taken within our limits at some future date. A specimen was shot at Lake Umbagog, New Hampshire, October 2, 1896, by Mr. Charles Douglass. (Cf. Brewster, The Auk, April, 1897, p. 207).

Genus ANSER Brisson.

11. (171a). Anser albifrons gambeli (*Hartl.*). American White-fronted Goose.

Given by Mr. Boardman as accidental at Grand Menan, New Brunswick.

Subfamily CYGNINÆ. Swans.

Genus OLOR Wagler.

12. (180). Olor columbianus (*Ord*). Whistling Swan.

A specimen of this bird is said to have been taken near the mouth of the Kennebec River, at Brick Island, November, 1881, by William Williams, and it is recorded by Smith in Forest and Stream, Vol. 20, p. 125. While the identification of this specimen was probably correct, still the fact that it was neither preserved nor seen by an Ornithologist who was acquainted with the species in question, makes the record open to some slight doubt, and consequently the species is cited here.

Family ARDEIDÆ Herons, Bitterns, etc.

Genus ARDEA Linnæus.

Subgenus GARZETTA Kaup.

13. (197). Ardea candidissima *Gmel.* Snowy Heron.

Taken by Mr. Boardman at Grand Menan, where it was an accidental visitor from the south.

Family RALLIDÆ. Rails, Gallinules, and Coots.

Genus PORZANA Vieillot.

Subgenus CRECISCUS Cabanis.

14. (216). Porzana jamaicensis (*Gmel.*). Black Rail.

A Rail, probably of this species, was shot at Scarborough, October 4, 1881, but unfortunately was not preserved. For a record of this see Smith's List of the Birds of Maine, Forest and Stream, Vol. 20, p. 124.

Family RECURVIROSTRIDÆ. Avocets and Stilts.
Genus HIMANTOPUS Brisson.

15. (226). Himantopus mexicanus (*Müll.*). Black-necked Stilt.

Given in Smith's List of the Birds of Maine, Forest and Stream, Vol. 20, p. 124, as being taken by Mr. Boardman near the mouth of the St. Croix River. In a list received from Mr. Boardman he states that this specimen was taken just over the boundary in New Brunswick, so consequently it cannot be given a place in the general list. It is a purely accidental visitant.

Family SCOLOPACIDÆ. Snipes, Sandpipers, etc.
Genus MACRORHAMPHUS Leach.

16. (232). Macrorhamphus scolopaceus (*Say*). Long-billed Dowitcher.

In Brown's Catalogue of the Birds of Portland, page 26, under *M. griseus*, he says: "I am confident that the western race, *scolopaceus*, visits us occasionally, although I have never met with it myself. Supposed examples of this form have twice been sent me for identification, in both instances, unfortunately, during my absence from town."

Genus EREUNETES Illiger.

17. (247). Ereunetes occidentalis *Lawr.* Western Sandpiper.
This species is of quite frequent occurrence on the Atlantic coast, and may possibly be looked for in this state.

Family CHARADRIIDÆ. Plovers.
Genus ÆGIALITIS Boie.
Subgenus OCHTHODROMUS Reichenbach.

18. (280). Ægialitis wilsonia (*Ord*). Wilson's Plover.
Given in the A. O. U. Check List, page 102, as being casual north to Nova Scotia.

Family HÆMATOPODIDÆ. Oyster-catchers.

Genus HÆMATOPUS Linnæus.

19. (286). Hæmatopus palliatus *Temm.* American Oyster-catcher.

Accidental, a specimen having been taken near Eastport, Washington County, by Mr. Boardman. (Cf. Smith, Forest and Stream, Vol. 20, p. 45). I am informed by Mr. Boardman that this bird was really taken on Canadian soil, and consequently is not a bird of the state.

Family TETRAONIDÆ. Grouse, Partridges, etc.

Genus BONASA Stephens.

20. (300). Bonasa umbellus (*Linn.*). Ruffed Grouse.

While this species has been given in every previous state and county list published, yet there is no positive proof that specimens have actually been taken in the state. Nearly all records of this bird are referable beyond doubt to the Canadian Ruffed Grouse. It has been deemed advisable to refer this species to the hypothetical list until satisfactory proof of its presence in the state has been adduced.

Family FRINGILLIDÆ. Finches, Sparrows, etc.

Genus GUIRACA Swainson.

21. (597). Guiraca cærulea (*Linn.*). Blue Grosbeak.

In New England Bird Life the author states that this species was recorded as being found near Calais by Mr. Boardman (Proc. Bost. Soc. Nat. Hist., 9, 1862, p. 127), and also that it is included in Herrick's "Catalogue of the Birds of Grand Menan." It is very probable, in view of later developments, that both of these instances refer to one specimen which was taken at Grand

---

The American Barn Owl, *Strix pratincola Bonap.*, has not a particle of right to be rated as a Maine bird. Smith's List cited it, apparently upon the evidence of the notorious "Falmouth specimen."

The Boat-tailed Grackle, *Quiscalus major Vieill.*, is given in Smith's List as being seen by him at Second Lake, Washington County, but personally I must confess that I doubt this record, and until a specimen has been taken within our limits, I must decline to recognize this species as a bird of Maine or even New England. It is apparently not even entitled to a place in the hypothetical list.

Menan, and accordingly the species is not entitled to a place in the
state. In a letter received from Mr. Boardman he gives this as a
Grand Menan bird only, two specimens having been taken there.

Family VIREONIDÆ. Vireos.
Genus VIREO Vieillot.

22. (631). Vireo noveboracensis (*Gmel.*). White-eyed Vireo.
The White-eyed Vireo was given in Smith's List, but its claim to
a place in our list is not based upon satisfactory evidence. Being
essentially a bird of the Carolina fauna its occurrence is very
doubtful.

Family MNIOTILTIDÆ. Wood Warblers.
Genus HELMITHERUS Rafinesque.

23. (639). Helmitherus vermivorus (*Gmel.*). Worm-eating
Warbler.
The sole claim of this species to a place in the list is based upon
a specimen from Maine recorded by A. E. Verrill. (Cf. Verrill,
Proc. Essex Inst., Vol. 3, p. 156). As I have not seen this publication, I am unable to state the grounds upon which he has
recorded the bird's presence in the state, but regard its occurrence
as very doubtful.

Genus HELMINTHOPHILA Ridgway.

24. (646). Helminthophila celata (*Say*). Orange-crowned
Warbler.
Ascribed to Maine by Audubon, but as he mentions the species
as breeding in eastern Maine it seems probable that his record is
the result of a misapprehension. There is certainly no modern
record of its occurrence in the state.

Family TROGLODYTIDÆ. Wrens, Thrushes, etc.
Genus CISTOTHORUS Cabanis.

25. (724). Cistothorus stellaris (*Licht.*). Short-billed Marsh
Wren.
This species is credited to Penobscot County in Smith's List.
(Cf. Smith, Forest and Stream, Vol. 19. p. 445). He here states
that the nests and eggs have been taken in Penobscot County. I

have given this subject a thorough investigation, and am unable to obtain any authentic data regarding these reported nests and eggs. One so-called Marsh Wren's nest I have seen has proved to be that of the field mouse, and quite different from genuine nests of the species. There have been no birds of this species actually taken within our limits, and consequently I am obliged to relegate this species to the hypotbetical list. Nevertheless I will state that personally I have good grounds for believing that these birds occur with us. I have seen birds in a marsh, near Bangor, which I am very sure were Marsh Wrens.

Family TURDIDÆ. Thrushes, Solitaires, Stonechats, Blue-birds, etc.

Genus TURDUS Linnæus.

26. (757 a). Turdus aliciæ bicknelli (*Rigdw.*). Bicknell's Thrush.

This species undoubtedly occurs as a migrant, and in all probability it will ultimately be found breeding on some of our higher mountain ranges, as it is already known to breed on the White Mountains in New Hampshire.

Genus SAXICOLA Bechstein.

27. (765). Saxicola œuanthe (*Linn.*). Wheatear.

There are no cases of the occurrence of this species on New England soil, all records to the contrary notwithstanding. These records have all been based upon specimens taken by Mr. Board-man, and in a recent letter he informs me that one of these birds was taken at Grand Menan, New Brunswick, while the other was taken August 25th, 1879, on Indian Island, near Eastport, but in New Brunswick. These are the only instances I am aware of where this bird has been found near our boundaries.

SUMMARY.

The number of species given in the list as positively occurring within our limits is 320. Of these I have included among the per-manent residents some 26 species, two of these being introduced and naturalized. The summer residents number 114. The species which occur chiefly or entirely as migrants are 74 in number. The winter residents and winter visitors of fairly regular occurrence

include 39 species. The accidental or casual visitants and stragglers include 65 species, while the remaining 2 species formerly
occurred within our limits but are now extinct in the state.

In the synopsis given below I have endeavored, with the evidence
at hand, to place each species under that heading which seemed to
most nearly represent its status in the state, while by appropriate
markings I have designated those which might be included under
other headings by another person. Persons disagreeing with my
grouping of any species will find in the text of the work, under that
species, data from which they are at liberty to draw their own
conclusions.

PERMANENT RESIDENTS.

The majority of the 26 species given here are resident,
but those designated by an asterisk are chiefly or entirely
~~entirely~~ confined to the Canadian fauna during the breeding season.
It is highly probable that in the case of some species the same
individual birds do not remain in one locality throughout the year, but
birds which summer here go south for the winter, and are replaced
during that season by individuals which have summered north of
our limits.

*Black Guillemot, *American Herring Gull, *Leach's Petrel, Bob-
White ?, *Canada Grouse, Canadian Ruffed Grouse, American
Long-eared Owl, Short-eared Owl, Barred Owl, *Saw-whet Owl,
Screech Owl, Great Horned Owl, Hairy Woodpecker, Downy Wood-
pecker, Pileated Woodpecker, Blue Jay, *Canada Jay, *Northern
Raven, American Crow, American Crossbill, White-winged Cross-
bill, White-breasted Nuthatch, *Red-breasted Nuthatch, Chickadee.

INTRODUCED SPECIES—Domestic Pigeon, English Sparrow.

SUMMER RESIDENTS.

The species which occur in greatest numbers as summer residents,
include 114 birds, some of which might have been equally well
included under one of the other headings. Those designated by an
asterisk have been known to occur in winter, though usually rare
at this season.

Pied-billed Grebe, *Loon, Laughing Gull, Common Tern, Arctic
Tern, *Black Duck, Wood Duck, American Bittern, Least Bittern,
Great Blue Heron, Green Heron, Black-crowned Night Heron, Virginia
Rail, Sora, American Woodcock, Bartramian Sandpiper, Spotted Sand-

piper, Piping Plover, Mourning Dove, Marsh Hawk, Sharp-shinned
Hawk, Cooper's Hawk, *Red-tailed Hawk, Red-shouldered Hawk,
Broad-winged Hawk, *Bald Eagle, American Sparrow Hawk,
American Osprey, Yellow-billed Cuckoo, Black-billed Cuckoo,
Belted Kingfisher, Yellow-bellied Sapsucker, Red-headed Wood-
pecker, Flicker, Whip-poor-will, Nighthawk, Chimney Swift, Ruby-
throated Hummingbird, Kingbird, Crested Flycatcher, Phœbe,
Olive-sided Flycatcher, Wood Pewee, Yellow-bellied Flycatcher,
Alder Flycatcher, Least Flycatcher, Bobolink, Cowbird, Red-
winged Blackbird, Meadow Lark, Baltimore Oriole, Bronzed
Grackle, *Purple Finch, *American Goldfinch, Vesper Sparrow,
Savanna Sparrow, Sharp-tailed Sparrow, Acadian Sharp-tailed
Sparrow, White-throated Sparrow, Chipping Sparrow, Field Spar-
row, Slate-colored Junco, Song Sparrow, Swamp Sparrow, Towhee,
Rose-breasted Grosbeak, Indigo Bunting, Scarlet Tanager, Purple
Martin, Cliff Swallow, Barn Swallow, Tree Swallow, Bank Swallow,
*Cedar Waxwing, Loggerhead Shrike, Red-eyed Vireo, Philadel-
phia Vireo, Warbling Vireo, Yellow-throated Vireo, Blue-headed
Vireo, Black and White Warbler, Nashville Warbler, Tennessee
Warbler, Northern Parula Warbler, Cape May Warbler, Yellow
Warbler, Black-throated Blue Warbler, Myrtle Warbler, Magnolia
Warbler, Chestnut-sided Warbler, Bay-breasted Warbler, Black-
burnian Warbler, Black-throated Green Warbler, Pine Warbler,
Yellow Palm Warbler, Oven-bird, Water-Thrush, Mourning Warbler,
Maryland Yellow-throat, Wilson's Warbler, Canadian Warbler,
American Redstart, Catbird, Brown Thrasher, House Wren, Winter
Wren, Brown Creeper, *Golden-crowned Kinglet, Wood Thrush
(very rare), Wilson's Thrush, Olive-backed Thrush, Hermit Thrush,
American Robin, Bluebird.

MIGRANTS OR TRANSIENT VISITORS.

I have here included some 74 species whose status seems
most nearly represented by this heading. Those designated
by an asterisk are summer residents to a greater or lesser
extent. Those marked with a dagger (†) are known to have
occurred in winter, some being regular winter residents in limited
numbers, while others are very rarely observed at this season.

* † Horned Grebe, † Red-throated Loon, Pomarine Jaeger,
Parasitic Jaeger, Long-tailed Jaeger, Ring-billed Gull,

†Bonaparte's Gull, Caspian Tern, *Roseate Tern, Greater Shearwater, Sooty Shearwater, †Gannet, * †American Merganser, * †Red-breasted Merganser, *Hooded Merganser, †Mallard, Bald-pate, Green-winged Teal, *Blue-winged Teal, Shoveller, Pintail, *Redhead, American Scaup Duck, Lesser Scaup Duck, *Ring-necked Duck, * †American Golden-eye, * †Buffle-head, *Ruddy Duck, Lesser Snow Goose, Canada Goose, Brant, *Yellow Rail, Florida Gallinule, American Coot, *Red Phalarope, Northern Phalarope, *Wilson's Snipe, Dowitcher, Stilt Sandpiper, Knot, Pectoral Sandpiper, White-rumped Sandpiper, Baird's Sandpiper, *Least Sandpiper, Red-backed Sandpiper, Semipalmated Sandpiper, Sanderling, Hudsonian Godwit, Greater Yellow-legs, Yellow-legs, *Solitary Sandpiper, Willet, Hudsonian Curlew, Eskimo Curlew, Black-bellied Plover, American Golden Plover, Killdeer, *Semi-palmated Plover, Turnstone, Passenger Pigeon, *Duck Hawk, *Pigeon Hawk, *Rusty Blackbird, *†Pine Siskin, Ipswich Sparrow, Nelson's Sparrow, White-crowned Sparrow, †Tree Sparrow, Lin-coln's Sparrow, Fox Sparrow, *Black Poll Warbler, American Pipit, *Ruby-crowned Kinglet, Gray-cheeked Thrush.

WINTER RESIDENTS OR WINTER VISITORS.

Under this heading I have placed 39 species. Those preceded by an asterisk are known to occur throughout the entire year, although some of these, notably the Scoters, do not breed within our limits.

Holbœll's Grebe, * Puffin, Murre, Brünnich's Murre, Razor-billed Auk, Dovekie, Kittiwake, Glaucous Gull, Iceland Gull, Great Black-backed Gull, Cormorant, * Double-crested Cormorant, Barrow's Golden-eye, * Old Squaw, Harlequin Duck, Northern Eider, * American Eider, King Eider, * American Scoter, * White-winged Scoter, * Surf Scoter, Purple Sandpiper, * American Gos-hawk, American Rough-legged Hawk, Great Gray Owl, Richard-son's Owl, Snowy Owl, American Hawk Owl, * Arctic Three-toed Woodpecker, * American Three-toed Woodpecker, Horned Lark, Prairie Horned Lark, * Pine Grosbeak, * Redpoll, Snowflake, Lap-land Longspur, Bohemian Waxwing, Northern Shrike, *Hudsonian Chickadee.

ACCIDENTAL OR CASUAL VISITANTS AND STRAGGLERS.

It seems very hard to draw any definite line between birds which are accidental or casual and those which are stragglers. The 65 species given here are either not regular in their occurrence, purely accidental, or, in the case of Wilson's Petrel and one or two others, birds which migrate northwards at the close of the breeding season for some unknown reason. Birds placed here, with the evidence now at hand, may ultimately be proved of regular occurrence, and they can then be assigned to some other group.

Tufted Puffin, Herring Gull, Sabine's Gull, Gull-billed Tern, Least Tern, Sooty Tern, Black Tern, Black Skimmer, Pintado Petrel, Wilson's Petrel, American White Pelican, Gadwall, Canvas-back, Greater Snow Goose, Hutchin's Goose, Wood Ibis, American Egret, Little Blue Heron, King Rail, Clapper Rail, Corn Crake, Purple Gallinule, Wilson's Phalarope, American Avocet, Curlew Sandpiper, Marbled Godwit, Ruff, Buff-breasted Sandpiper, Long-billed Curlew, Belted Piping Plover, Willow Ptarmigan, Turkey Vulture, Black Vulture, Swainson's Hawk, Golden Eagle, White Gyrfalcon, Gray Gyrfalcon, Gyrfalcon, Black Gyrfalcon, Arctic Horned Owl, Dusky Horned Owl, Arkansas Kingbird, Starling, Yellow-headed Blackbird, Orchard Oriole, Bullock's Oriole, Evening Grosbeak, Amadina rubronigra (escaped cage-bird), Hoary Redpoll, Holboell's Redpoll, Greater Redpoll, Chestnut-collared Longspur, Grasshopper Sparrow, Seaside Sparrow, Cardinal, Dickcissel, Louisiana Tanager, Summer Tanager, Prothonotary Warbler, Louisiana Water-Thrush, Connecticut Warbler (probably rare migrant), Yellow-breasted Chat, Mockingbird, Carolina Wren, Blue-gray Gnatcatcher.

SPECIES NOW EXTINCT IN THE STATE.

The Great Auk formerly occurred as a winter visitant but it is now extinct. The Wild Turkey formerly occurred in the state, probably being a permanent resident. It has not been noted here for many years.

10

.

FAUNAL AREAS.

Zoogeography, or the geographical distribution of species, is a comparatively new science, but, nevertheless, one destined to prove of the utmost economic importance to the agriculturist. Plants and animals are naturally found in certain climates where the conditions of temperature and surroundings are congenial to them. Some species have the power of adapting themselves to circumstances and are found in many diverse climates, while others are exclusively confined to certain regions over which the same conditions prevail, and need not be looked for elsewhere. These latter are said to be indigenous to a life area, and by aid of these species we are enabled to divide the country into an ultimate number of life areas which are termed faunæ. When by study and observation of a certain plant or animal in various localities, we have finally arrived at the conclusion that wherever we have found that particular species the conditions of climate, surroundings, etc., are uniform, we may then safely say, on hearing from some outside observer that this same species is found in his locality, just what the climate and surroundings of that locality are, without ever having seen it. Of course we are always open to error due to the fact that this species may be able to adapt itself to other conditions, but from a close and long-continued study of certain North American plants and animals, it has been almost positively demonstrated that they are exclusively confined to regions over which similar conditions prevail. In some cases the primary life areas are characterized by the prevalence over them of entire families or genera, while in the case of the minor or faunal areas the prevalence of certain species or subspecies and, equally important, the absence of others characteristic of other areas, are links in the chain of evidence by which we are enabled to map the limits of these divisions.

The change from one area to another is not at all abrupt, but instead as we near their common boundary we find species common to both occurring on the same grounds. In such cases, the prevalence of the species of one of the faunæ will result in determining the area to which that region belongs.

It is of the utmost importance to note the fact that these life areas do not regularly blend with one another, but the points of their intergradation may be compared to the meeting of the water and land along the irregular, indented coast of Maine. The irregularity of these life areas and the mapping out of their various spurs and projections are of the utmost importance to the farmer. By utilizing the northern projections of a more southern fauna he is able to grow its indigenous plants just so much nearer to a northern market, while some dozens of miles to the eastward his neighbor may be utilizing a southward extension of a colder life area to grow boreal plants so much nearer to a southern market. In general the southern extensions of the colder areas will be found along the higher mountain ranges, while the northern branches of the warm areas are in the lowlands.

With these explanations we will proceed to an enumeration of the various primary and secondary areas. The entire world has been divided into eight primary life areas, termed Realms, as follows:

(1) The Arctic Realm extends across the northern continents, reaching from the northern limit of forest growth to the pole. It is remarkable for the paucity and specific identity of the forms of life occurring throughout it.

(2) The North Temperate Realm extends from the northern limit of forest growth south to the palm tree belt.

(3) The American Tropical Realm includes tropical America.

(4) The Indo-African Realm consists of all Africa, except the northern portion, and tropical Asia with its islands.

(5) The South American Temperate Realm includes temperate South America.

(6) The Australian Realm embraces Australia and the islands of Oceanica adjacent thereto.

(7) The Lemurian Realm is confined to the island of Madagascar.

(8) The Antarctic Realm occupies the same position in the south as does the Arctic in the north, and the species inhabiting it are likewise few and of general distribution. The birds are mainly pelagic.

All of extreme northern North America is within the Arctic Realm, south of this comes the North Temperate which extends quite to the southern boundary of the United States, except in Florida and Texas where the American Tropical Realm enters their extreme southern portions.

Owing to lack of space, I will not enter into a discussion of the minor life areas except such as concern Maine directly. Any one who may wish to enter into a thorough investigation of this subject will find interesting articles on it as follows : The Geographical Distribution of North American Mammals, J. A. Allen, Bull. Am. Mus. Nat. Hist., Vol. 4, pp. 199-243. The Geogr. and Geol. Distribution of North American Animals. The Origin and Distribution of North American Birds, J. A. Allen, The Auk, Vol. 10, pp. 99-150. Various reports of the U. S. Department of Agriculture, Division of Ornithology and Mammology, contain interesting articles by Dr. Merriam who is well known as an authority on this subject.

The North Temperate Realm is divided into regions of which the North American Temperate Region alone concerns us. This in turn is divided into two subregions, the Cold Temperate and the Warm Temperate. The Cold Temperate Subregion is divided into four faunæ of which one, the Canadian, enters Maine. We have here one of our faunæ traced from its fountain head down through the classification.

The Warm Temperate is divided into two provinces, a Humid or Eastern Province and an Arid or Western Province. The Humid Province is divided into the Appalachian and Austroriparian Subprovinces, the former of which concerns us. This is divided into three faunæ, the northern of which is named the Alleghanian, and which enters our state in the southwestern part. We have then the Cold Temperate and Warm Temperate Subregions, as represented by the Canadian and Alleghanian Faunæ, meeting in our state. Under such condition one would expect to find a very interesting commingling of the species common to each, and such is the case.

In mapping out the Canadian Fauna I have used various characteristic trees, birds and animals, as aids in determining its southern limits. The forests of fir and spruce, indicate that the regions where they predominate are Canadian in character. The

Canada Porcupine, Northern Hare, Red Squirrel, and Jumping Mouse are characteristic mammals. A partial list of the birds is found in the table below.

BIRDS OF THE CANADIAN FAUNA.

Black Guillemot, American Herring Gull, Leach's Petrel, Red-breasted Merganser,, American Goshawk, Olive-sided Fly-catcher, Yellow-bellied Flycatcher, Canada Jay, Northern Raven, Rusty Grackle, Pine Siskin, Acadian Sharp-tailed Sparrow, White-throated Sparrow, Slate-colored Junco, Red-breasted Nuthatch, Olive-backed Thrush, Golden-crowned Kinglet, Water-Thrush, Brown Creeper, Winter Wren, Myrtle Warbler, Black-throated Blue Warbler, Bay-breasted Warbler, Black Poll Warbler, Wilson's Warbler, Magnolia Warbler, Alder Flycatcher.

The above is merely a partial list of the birds which distinguish the limits of this fauna. Some of these occur in slight numbers in the Alleghanian while others are extreme Canadian types and occur well within its limits. However they may all be regarded as fairly distinctive.

The Alleghanian Fauna is characterized by such trees as the pine and oak. The birds are given below.

Least Bittern, Green Heron, Mourning Dove, Meadow Lark, Yellow-billed Cuckoo, Field Sparrow, Sharp-tailed Sparrow, Wood Thrush, Towhee, Brown Thrasher, House Wren, Bob-white.

The birds cited above may be considered fairly typical of their respective faunae, and the prevalence of the species of one over those of the other will settle to which fauna a given locality belongs.

Previous observers have assigned the dividing line between our faunae to a somewhat indefinite locality near Mount Desert Island. Beginning here, the Alleghanian Fauna has been stated to include the territory south of the line of mountains which run in a south-westerly direction across the state. Part of this is wrong in view of information of which I am now possessed.

We may safely assign to the Canadian Fauna the entire granite-ridged, spruce-covered sections of the coast. The Laurentian Hills with their outspurs present features which are in strong contrast to those of the southwestern part of the state. The southern limit of growth of the low, stunted spruces of the coast is coincident with the distribution of the majority of Canadian birds, although many are not found quite so far southwards.

In a recent article regarding the "Sharp-tailed Finches of Maine" Mr. A. H. Norton speaks of the habitat of the Acadian Sharp-tail as follows : "North of Scarboro, beginning with Cape Elizabeth, its eastern boundary, the coast presents an uneven or hilly face of rocks, indented with numerous coves and bays, studded with dry ledgy islands. Between the hills are innumerable arms of the sea often extending as "tide-rivers" or fjords several miles inland, bordered by narrow swales. Coincident with these features is the low spruce woods, so conspicuous a feature of the Maine coast, so characteristic of the scanty soiled granite ridges, and the fog drenched coast of the northeast. Very different in appearance are the broad marshes of Scarboro and western Maine, backed by soil-clad verdant slopes, with pine and hard woods replacing the spruce." (Cf. Norton, Proc. Port. Soc. Nat. Hist., Vol. 2, pp. 100-101).

In my judgment he has here outlined the dividing line between the two faunæ, and the Canadian thus extends along the coast to Cape Elizabeth. A few miles back in the interior these Canadian features cease, and we will find ourselves in the midst of Alleghanian surroundings. These latter extend eastward in the interior into Lincoln County, where the two regions seemingly meet, as is evidenced by the pine trees and spruces being about equal in numbers. To the northward the Alleghanian surroundings probably predominate till the line of mountains which runs southwesterly across the state is reached. In the absence of specific proof regarding this point, I have been obliged to accept the evidence of previous writers on the subject and follow their conclusions.

All of the region to the eastward of the Penobscot River, together with that north of the before-mentioned chain of mountains, is Canadian in character. Of this there can be no doubt. We may therefore say that Aroostook, Franklin, Hancock, Penobscot, Piscataquis, Somerset, and Washington Counties are Canadian. To these may be added, provisionally, Knox and Waldo. A narrow Canadian strip extends along the coast through Lincoln and Sagadahoc into Cumberland. The northern parts of Androscoggin, Kennebec, and Oxford may be also included in this fauna.

The Alleghanian includes all of York County and such parts of Androscoggin, Cumberland, Kennebec, Lincoln, Oxford, and Sagadahoc as have not been previously designated as Canadian.

With the information at hand this is the best I am able to do in re-arranging our faunal lines. It is highly probable that isolated areas of one fauna will be found to occur well within and entirely surrounded by the other. These isolated areas will ultimately be mapped out with a precision which cannot now be attempted. During the past month I have obtained information which has changed conclusions I formerly held regarding the limitations of these areas.

From personal observation I am able to state that Hancock County is purely Canadian in characters, and this is likewise true with Penobscot, where Canadian characters slightly predominate. Such parts of Waldo as I have been over are likewise Canadian, although I have not been in the southwestern part. The conclusions arrived at regarding the other counties are derived from the combined observations of others and myself.

BIBLIOGRAPHY.

I have given below a partial list of various books and publications which contain articles relating more or less directly to the Ornithology or Oölogy of the state. The titles of many may not be exactly or correctly quoted here, owing to the fact that I have seen but a very small portion of the publications cited. I doubt not but that most of them will be recognized under the titles given them. The list does not claim to be either complete or exact, but is given to serve as a slight aid to persons who wish to go more completely into the literature relating to our birds. Such publications as Coues's Key, Ridgway's Manual, and others relating to the birds of North America at large are cited because of their general bearing on the subject, and for this same reason a few of the numerous periodicals are also given. Notes regarding many of the publications cited have been taken from Stearn's New England Bird Life.

1832. Williamson, William D. The History of the State of Maine (etc.). For notes on birds see pp- 140-150 of Vol. 1.

1861. Holmes, E. Zoology of Maine. See Sixth Annual Report of Secretary of Maine Board of Agriculture, pp. 113-122. About 193 species nominally listed.

1862. Catalogue of the Birds of Maine. In Proceedings of the Portland Society of Natural History, Vol. 1, pp. 66-71. About 230 species, some of which were reported without the slightest evidence of their occurrence. List nominal.

1862. Boardman, George A. Catalogue of the Birds Found in the Vicinity of Calais, Maine, and about the Islands of the Bay of Fundy. Proceedings of the Boston Society of Natural History, Vol. 9, pp. 122-132. Two hundred and thirty-one species with annotations.

1862. Holmes, E. Birds of Maine. Addenda. Second Annual Report of Natural History and Geology of Maine, p. 118.

1862. Verrill, A. E. Catalogue of the Birds Found at Norway, Oxford County, Maine. Proceedings of the Essex Institute, Vol. 3, pp. 136-160. Annotated list of 159 species and also a list of 107 Maine birds not seen at Norway.

1863. Samuels, Edward A. Mammology and Ornithology of New England. Report of United States Commissioner of Agriculture, 1863, pp. 265-286. Of little importance regarding Maine species.

1863. Verrill, A. E. Additions to the Catalogue of the Birds Found in the Vicinity of Calais, Maine, and About the Islands of the Bay of Fundy. Proceedings of the Boston Society of Natural History, Vol. 9, pp. 233-234. Twelve species added to Boardman's 1862 list.

1865. Hamlin, Charles E. Catalogue of the Birds Found in the Vicinity of Waterville, Kennebec County. Tenth Annual Report of the Secretary of the Maine Board of Agriculture, 1865, pp. 168-173.

1867. Samuels, E. A. Ornithology and Oölogy of New England. Some Maine birds are referred to in this publication.

1867. Wyman, Jeffries. An Account of Some KJœkkenmœddings or Shell-heaps in Maine and Massachusetts. American Naturalist, Vol. 1, 1867, pp. 561-584.

1868. Coues, Elliott. A List of the Birds of New England.

1869. Boardman, George A. Breeding of Rare Birds (at Milltown). American Naturalist, Vol. 3, 1869, p. 222.

1869. Boardman, George A. The Black Vulture in Maine. American Naturalist, Vol. 3, 1869, p. 498.

1871. Boardman, Geo. A. Ornithological Notes from Maine. American Naturalist, Vol. 5, 1871, p. 662.

1872. Maynard, C. J. A Catalogue of the Birds of Coos County New Hampshire and Oxford County, Maine. With Notes by William Brewster. Proceedings of the Boston Society of Natural History, 1871, pp. 356-385. This is an annotated list of 164 species.

1873. Brewer, Dr. T. M. Catalogue of the Birds of New England. Proceedings of the Boston Society of Natural History, Vol. 17, 1875, pp. 436-454. Contains notes relating to the status of most species.

1875. Brown, Nathan C. Ornithological Notes from Portland, Maine. Rod and Gun, Vol. 6, 1875, p. 81.

1875. Flagg, Wilson. Birds and Seasons of New England.

1877. Brown, Nathan C. Notes on Birds New to the Fauna of Maine. Bulletin of the Nuttall Ornithological Club, Vol. 2, January 1877, pp. 27-28. Five species given.

1877. · Minot, H. D. The Land-Birds and Game-Birds of New
      England, with Descriptions of the Birds, Their Nests and
      Eggs, Their Habits and Notes.
1878. Brewer, Dr. T. M. Notes on Certain Species of New
      England Birds, with additions to his Catalogue of the
      Birds of New England. Proceedings of the Boston
      Society of Natural History, Vol. 19, pp. 301-309.
1879. Brown, Nathan C. Notes on a Few Birds Occurring in
      the Vicinity of Portland, Maine. Bulletin of the Nuttall
      Ornithological Club, Vol. 4, 1879, p. 106.
1879. Boardman, Geo. A. Southern Birds Down East. Forest
      and Stream, Vol. 13, p. 605. Notes the occurrence of
      the Black Skimmer and Black Vulture in Maine.
1881. Stearns, W. A. New England Bird Life, a Manual of
      Ornithology. Edited by Dr. Coues. Part 1. Oscines,
      Singing Birds. Lee and Shephard, Boston.
1882. Brown, Nathan Clifford. A Catalogue of the Birds Known
      to Occur in the Vicinity of Portland, Me., Especially in
      the Townships of Falmouth, Deering, Westbrook, Cape
      Elizabeth and Scarborough, Briefly Annotated. Proceed-
      ings of the Portland Society of Natural History, Decem-
      ber 4, 1882. 248 species are given here with more or
      less complete annotations.
1882-83. Smith, Everett. The Birds of Maine, with Annotations
      of Their Comparative Abundance, Dates of Migration,
      Breeding Habits, etc. Forest and Stream, Vol. 19, 1882,
      Nos. 22-26, and Vol. 20, 1883, Nos. 1-7 and 10-13.
      Total number of species given here is 303 but some are
      included on insufficient evidence.
1883. Stearns, W. A. New England Bird Life, etc. Part 2,
      Birds of Prey, Game and Water Birds. Lee and Shep-
      hard, Boston. Many records of birds from Maine are
      cited here.
1889. Brown, Nathan Clifford. Supplementary Notes on the
      Birds of Portland and Vicinity. Proceedings of the Port-
      land Society of Natural History, Vol. 2, part 1, p. 37.
      Adds 8 species to his previous list, and gives additional
      notes on 33 species.

1889. Davie, Oliver. Nests and Eggs of North American Birds. Hann and Adair, Columbus. References are made here to the nesting of certain species in Maine.

1891. Walter, Herbert E. The Birds of Androscoggin County. Notes on the Perching Birds of Androscoggin County, Supplemented by a Catalogue of Other Species, Excluding the Shore and Water Birds, also identified in the county. From the History of Androscoggin County. This is an annotated list of 95 species, while 38 more are given nominally. Eighteen other species are mentioned as being reported in the state but not yet observed in the above county.

1892. Bendire, Captain Charles. Life Histories of North American Birds; with Special Reference to their Breeding Habits and Eggs, with Twelve Lithographic Plates. Government Printing Office, Washington. Gallinaceous Birds—Raptores. Quotes Manly Hardy extensively regarding the habits of many Maine birds.

1895. Minot, Henry Davis. The Land-Birds and Game-Birds of New England. Second Edition, edited by William Brewster. The Riverside Press, Cambridge. Many Maine birds are recorded here.

1895. Chapman, Frank M. Handbook of Birds of Eastern North America, etc. Third Edition. New York, D. Appleton & Co. The occurrence of certain species in Maine is cited here.

1895. Bendire, Captain and Brevet Major Charles. Life Histories of North American Birds from the Parrots to the Grackles, with Special Reference to their Breeding Habits and Eggs, with Seven Lithographic Plates. Washington, Government Printing Office. Quotes Mr. Hardy regarding Maine birds.

1895. American Ornithologists' Union. The A. O. U. Check-List of North American Birds. Second Edition. Cambridge, Mass. Refers to many birds as inhabiting Maine.

1896. Ridgway, Robert. A Manual of North American Birds. Illustrated by 464 Outline Drawings of the Generic Characters. Second Edition. J. Lippincott Company, Philadelphia.

1897. Norton, Arthur H. Sharp-tailed Finches of Maine. Remarks on their Relationship and Distribution. Proceedings of the Portland Society of Natural History, Vol. 2, March 15th, 1897, pp. 97-102.

——. Coues, Elliot. Key to North American Birds.

——. Chamberlain, Montague. Nuttall's Ornithology. A New and Revised Edition. Nuttall's Land, Game and Water Birds, Colored Plates and Many Illustrations. 2 Vols. Refers to certain Maine birds.

——. Capen's Oology of New England.

——. Maynard's Birds of Eastern North America.

——. Wilcox, ——, Common Land Birds of New England.

PERIODICALS IN WHICH REFERENCES TO MAINE BIRDS HAVE BEEN PUBLISHED.

The Auk, a Quarterly Journal of Ornithology, Published for the American Ornithologists' Union. New York, L. S. Foster. $3.00 per annum.

The Osprey, an Illustrated Monthly Magazine of Ornithology, Edited by Walter A. Johnson, associated with Dr. Elliot Coues. Published by the Osprey Company, Galesburg, Illinois. Subscription $1.00 yearly.

The Nidologist, Exponent of American Ornithology and Oölogy, Published Monthly with Illustrations, by Henry Reed Taylor, Alameda, California. Subscription $1.00 per year.

The Oölogist. Monthly. Published by Frank H. Lattin, Albion, New York. Subscription 50 cents per year.

The Museum, a Journal Devoted to Research in Natural Science. Published the Fifteenth of Each Month by Museum Publishing Company, Walter F. Webb, Manager, Albion, N. Y. $1.00 yearly.

Maine Sportsman, Published the First of Every Month. Herbert W. Rowe, Bangor. $1.00 per year. The official organ of the United Ornithologists of Maine.

ADDENDA.

Since the first forms of the list have gone to press, many addi-· tional notes of interest have been received from various sources. Mr. Everett Smith should have been given credit for furnishing a number of notes on the occurrence of certain species in various parts of the state. Mr. A. H. Norton has recently furnished a partial list of birds observed in Lincoln County. Such notes as were received too late for insertion in their proper places in the list are given here.

Colymbus holbœllii (*Reinh.*). Holbœll's Grebe.
"Winter resident, Knox County" (Norton).

Colymbus auritus *Linn.* Horned Grebe.
"Winter resident, Knox County" (Norton).

Cepphus grylle (*Linn.*). Black Guillemot.
"Lincoln County, breeding in fair numbers in 1895" (Norton).

Uria lomvia (*Linn.*). Brünnich's Murre.
"Two specimens found dead in the ice in Bridgton and Otisfield" (Mead).

Stercorarius pomarinus (*Temm.*). Pomarine Jaeger.
"Lincoln County, three seen on June 23rd and again on the 24th, 1895" (Norton).

Larus leucopterus *Faber.* Iceland Gull.
"Knox County in winter" (Norton).

Larus argentatus smithsonianus *Cones.* American Herring Gull.
"Lincoln County" (Norton).

Larus atricilla *Linn.* Laughing Gull.
"Lincoln County, about 14 birds breeding in June, 1895. I took an egg in Knox County in 1896" (Norton).

Larus philadelphia (*Ord*). Bonaparte's Gull.
"Lincoln County, observed in June, 1895" (Norton).

Sterna hirundo *Linn.* Common Tern.
"Lincoln County, still quite common and breeding in 1895" (Norton).

Stern paradis:ea *Brünn.* Arctic Tern.

"Lincoln County, still quite common and breeding in 1895" (Norton).

Hydrochelidon nigra surinamensis (*Gmel.*). Black Tern.

"Knox County, rare, credit is due Mr. Rackliff for taking a specimen in this county" (Norton).

Oceanodroma leucorhoa (*Vieill.*). Leach's Petrel.

"A few breeding in Lincoln County" (Norton).

Sula bassana (*Linn.*). Gannet.

"Lincoln County, one seen in June, 1895" (Norton).

Aythya vallisneria (*Wils.*). Canvas-back.

Prof. Wm. L. Powers writes that four specimens were shot near Gardiner, Kennebec County, in the fall of 1895. I have had the pleasure of viewing one of these which is at present in his collection.

Oidemia americana *Sw. and Rich.* American Scoter.

"Have seen a female which was shot near Gardiner in Kennebec County" (Knight).

Oidemia perspicillata (*Linn.*). Surf Scoter.

"Northern Cumberland, somewhat rare visitant" (Mead).

Ardea virescens *Linn.* Green Heron.

"Northern Cumberland, occasional" (Mead).

Nycticorax nycticorax nævius (*Bodd.*). Black-crowned Night Heron.

"Northern Cumberland, one specimen, young, in twenty-five years" (Mead).

Nyctea nyctea (*Linn.*). Snowy Owl.

"Waldo County, occasional" (Spratt).

Agelaius phœniceus (*Linn.*). Red-winged Blackbird.

"Waldo County, common summer resident" (Spratt).

Acanthis linaria rostrata (*Coues*). Greater Redpoll.

Mr. A. H. Norton writes: "I regret to note that I mailed you an uncorrected copy of my notes on the Redpolls. The year, 1895, should have been 1896." Accordingly all his notes which are quoted regarding this subspecies should be dated 1896.

Lanius borealis *Vieill.* Northern Shrike.

"Waldo County" (Spratt).

Dendroica blackburniæ (*Gmel.*). Blackburnian Warbler.

"Waldo County, rare" (Spratt).

Anthus pensilvanicus (*Lath.*). American Pipit.

"Was common at Wilson's Mills, Oxford County, in September and October, 1879" (Mead).

INDEX.

A.

II

B.

D.

G.

H.

I.

M.

O.

P.

Q.

R.

S.

12

U.

V.

W.

X.

Y.

ERRATA.

Page 15, No. 5, line 6, pond should read Pond.
" 26, No. 43, line 7, Fauna should read fauna.
" 27, No. 46, line 4, Fauna should read fauna.
" 28, No. 48, line 1, Anas boschas should read Anas boschas.
" 34, No. 68, line 3, eastwerd should read eastward.
" 37, No. 77, line 10, ant should read an.
" 47, No. 112, last line, Borsdman should read Boardman.
" 90, No. 215, line 2, varying should read varying.
" 94, No. 222, line 2, line 5, 1895 should read 1896.
" 97, No. 233, line 10, *cavdacutus* should read *caudacutus*.
" 98, No. 234, line 14, should read (Cf. Norton, Proc. Port. Soc. Nat. Hist., 1897
　　　　p. 100).
" 142, line 15, omit "entirely."